EAST-WEST TECHNOLOGY TRANSFER

I
CONTRIBUTION TO EASTERN GROWTH : AN ECONOMETRIC EVALUATION

by
Stanislaw Gomulka and Alec Nove

II
SURVEY OF SECTORAL CASE STUDIES

by
George D. Holliday

ORGANISATION FOR ECONOMIC CO-OPERATION AND DEVELOPMENT

Pursuant to article 1 of the Convention signed in Paris on 14th December, 1960, and which came into force on 30th September, 1961, the Organisation for Economic Co-operation and Development (OECD) shall promote policies designed:

- to achieve the highest sustainable economic growth and employment and a rising standard of living in Member countries, while maintaining financial stability, and thus to contribute to the development of the world economy;
- to contribute to sound economic expansion in Member as well as non-member countries in the process of economic development; and
- to contribute to the expansion of world trade on a multilateral, non-discriminatory basis in accordance with international obligations.

The Signatories of the Convention on the OECD are Austria, Belgium, Canada, Denmark, France, the Federal Republic of Germany, Greece, Iceland, Ireland, Italy, Luxembourg, the Netherlands, Norway, Portugal, Spain, Sweden, Switzerland, Turkey, the United Kingdom and the United States. The following countries acceded subsequently to this Convention (the dates are those on which the instruments of accession were deposited): Japan (28th April, 1964), Finland (28th January, 1969), Australia (7th June, 1971) and New Zealand (29th May, 1973).

The Socialist Federal Republic of Yugoslavia takes part in certain work of the OECD (agreement of 28th October, 1961).

Publié en français sous le titre:

TRANSFERT DE TECHNOLOGIE
ENTRE L'EST ET L'OUEST

I. SA CONTRIBUTION AU DÉVELOPPEMENT
DES PAYS DE L'EST:
ANALYSE ÉCONOMÉTRIQUE

II. EXAMEN D'ETUDES DE CAS SECTORIELLES

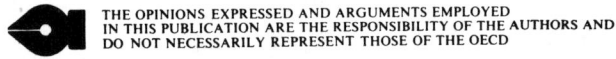

THE OPINIONS EXPRESSED AND ARGUMENTS EMPLOYED
IN THIS PUBLICATION ARE THE RESPONSIBILITY OF THE AUTHORS AND
DO NOT NECESSARILY REPRESENT THOSE OF THE OECD

© OECD, 1984
Application for permission to reproduce or translate
all or part of this publication should be made to:
Director of Information, OECD
2, rue André-Pascal, 75775 PARIS CEDEX 16, France.

While self-contained and entirely the responsibility of the authors, these studies were prepared within the context of a wider project concerning East-West Technology Transfer conducted under the auspices of the Committee for Scientific and Technological Policy.

It was composed of two distinct, though interrelated, phases. The first was exploratory in nature and delineated the field of inquiry. It was concluded with the publication entitled *Technology Transfer between East and West*.

The second seeks to analyse and assess two broad issues: the factors determining the assimilative capacity of the Eastern countries with regard to technology in general and Western technology in particular and the impact of Eastern imports of Western technology on East-West trade in technology.

The method of analysis consisted of country studies, analyses based on East-West and intra-CMEA trade flows, and on econometric and sectoral analyses.

Publication of this report has been authorised by the Secretary-General.

Also available

EAST-WEST TECHNOLOGY TRANSFER
Study of Poland 1971-1980, by Zbigniew Fallenbuchl (September 1983)
(92 83 01 1) ISBN 92-64-12484-5 200 pages   £11.00 US$22.00 F110.00

NORTH/SOUTH TECHNOLOGY TRANSFER. THE ADJUSTMENTS AHEAD. Analytical Studies. "Document" Series (March 1982)
(92 81 07 1) ISBN 92-64-12265-6 222 pages   £9.00 US$20.00 F90.00

NORTH/SOUTH TECHNOLOGY TRANSFER. THE ADJUSTMENTS AHEAD (March 1981)
(92 81 02 1) ISBN 92-64-12159-5 116 pages   £4.80 US$12.00 F48.00

SCIENCE AND TECHNOLOGY POLICY FOR THE 1980s (November 1981)
(92 81 05 1) ISBN 92-64-12254-0 168 pages   £6.20 US$13.75 F62.00

TECHNOLOGY TRANSFER BETWEEN EAST AND WEST, by Eugène Zaleski and Helgard Wienert (September 1980)
(92 80 02 1) ISBN 92-64-12125-0 436 pages   £22.00 US$50.00 F200.00

Prices charged at the OECD Publications Office.

THE OECD CATALOGUE OF PUBLICATIONS and supplements will be sent free of charge on request addressed either to OECD Publications Office, 2, rue André-Pascal, 75775 PARIS CEDEX 16, or to the OECD Sales Agent in your country.

TABLE OF CONTENTS

PREFACE 7

THE AUTHORS 8

Book I

ECONOMETRIC EVALUATION OF THE CONTRIBUTION OF WEST-EAST TECHNOLOGY TRANSFER TO THE EAST'S ECONOMIC GROWTH

by
Stanislaw Gomulka and Alec Nove

Part A: INTRODUCTION, THE MEASUREMENT PROBLEM AND PRELIMINARY EVIDENCE 11
1. Introduction 11
2. Technological Change and Output Growth: the Measurement Problem in the Short Run and Long Run 13
3. The Preliminary Statistics 16

Part B: THE WORK DONE AND THE RESULTS REPORTED: A SURVEY 18
1. *Green-Levine:* SOVMOD III, as reported in "Macroeconometric Evidence of the Value of Machinery Imports to the Soviet Union", Stanford Research Institute, 1977, and in "Soviet Machinery Imports", *Survey,* Spring 1978, Vol. 23, No. 2, pp. 112-126 18
2. *Weitzman:* "Technology Transfer to the USSR: an Econometric Analysis". *Toda:* "Technology Transfer to the USSR: the Marginal Productivity Differential and the Elasticity of Intercapital Substitution in Soviet Industry" 20
3. *Rosefielde:* "East-West Trade and Postwar Soviet Economic Growth: A Sectoral Production Function Approach" 22
4. *Gomulka:* "Inventive Activity, Diffusion and the Stages of Economic Growth", Aarhus Institute of Economics, 1971; "Import-led Growth: Theory and Estimation", 1976; "Growth and the Import of Technology: Poland 1971-1980", *Cambridge Journal of Economics,* March 1978 22
5. Summary 25

Part C: CRITICAL COMPARATIVE EXAMINATION OF THE MEASUREMENT METHODS 26

Part D: IDENTIFICATION OF SPECIFIC PROBLEMS OF MEASUREMENT ... 30
1. Limitations of Aggregate Analysis 30
2. The Contribution of an Imported Machine in a Shortage-type Economy ... 31
3. Complementary Soviet Inputs and Priorities 32
4. Distinction Between Embargo Effects in Short Run and Longer Run 32
5. The Quality of Data on Imported Western Capital 33

Part E: SYSTEMIC FACTORS IN LOW ECONOMIC EFFICIENCY UNDER SOVIET-TYPE CENTRAL PLANNING AND MANAGEMENT 35

Part F: THE FAILURE OF THE IMPORT-LED GROWTH STRATEGY 38

Part G: CONCLUSION 39

NOTES 41

REFERENCES 42

Appendix 1: A SAMPLE OF QUOTATIONS FROM SOVIET OFFICIAL SOURCES ON THE DEFICIENCIES OF SOVIET INNOVATIONAL PROCESSES 44

Appendix 2: A SELECTION OF DATA 49

Book II

**TRANSFER OF TECHNOLOGY FROM WEST TO EAST:
A SURVEY OF SECTORAL CASE STUDIES**

by

George D. Holliday

I. INTRODUCTION 55
   1. Assimilative Capacity 56
   2. Impact on Trade 57

II. CASE STUDIES 60
   1. Methodologies and Sources 62
   2. Level of Technology and Degree of Dependence on West 65
      a) Level of technology transferred 65
      b) Eastern dependence on Western technology 66
   3. Channels or Mechanisms of Technology Transfer 69
      a) Evolution of technology transfer mechanisms 69
      b) Effects of mechanisms on assimilation 70
   4. Hard Currency Constraints and Efforts to Expand Exports 72
      a) Hard currency expenditures 72
      b) Tying technology purchases to exports of resultant products 73
      c) Competitiveness of Eastern exporters 75
   5. Indicators of Assimilative Capacity 76
      a) Lead-times 76
      b) Operation after startup 77
      c) Diffusion 78
   6. Effects of Domestic Infrastructures and Institutions on Assimilative Capacity 79
      a) Factors which impede assimilation 79
      b) Factors which facilitate assimilation 80

III. CONCLUSIONS AND SUGGESTIONS FOR FUTURE CASE STUDIES 82
   1. Conclusions on Assimilative Capacity and Impact on Trade 82
   2. Suggested Methodologies and Sources for Future Case Studies 85
   3. Suggested Sectors for Future Case Studies 87

NOTES 89

IV. SELECTED BIBLIOGRAPHY 90

PREFACE

The literature surveyed in this publication attempts to assess the size of the potential and actual gains that have been made from technology transfers from the West by the USSR and Eastern Europe for particular investment projects, individual sectors and for the economies of these countries as a whole. Particular emphasis is given to the economic effects of Western machinery imports.

The econometric studies surveyed in Book I report estimates of the contribution of these imports to Soviet industrial growth ranging from zero to about one per cent per annum for the period 1960-1975. The gains by Eastern Europe, in terms of the contribution of Western machinery imports to industrial growth, are thought to have been greater than those for the USSR. The authors of Book I came to the view that science and technology transfer is important for both the USSR and Eastern Europe and that channels other than machinery imports may be particularly important, especially for the USSR.

In Book II of the publication, the focus is placed on studies which have looked at the West-East technology transfer process by industry or by project. The purpose of this survey is to highlight the findings of case studies on the assimilative capacities of the Soviet and East European economies and the impact of West-East transfers of technology on trade. The author of this Book II points out that generalisations from findings of the studies surveyed should be made with caution. Nevertheless, he concludes that Western technology has made a significant contribution to the technological advancement of the Eastern industrial branches studied, but that domestic infrastructures and institutions tend to inhibit rapid and efficient assimilation of Western technology and the export of competitive manufactured goods to the West.

THE AUTHORS

Stanislaw GOMULKA is Senior Lecturer at the London School of Economics. In 1980-1981, he was a fellow of the Netherlands Institute of Advanced Studies. Most of his publications are in the fields of economic growth and innovation, especially in reference to centrally planned economies. Among these are:

"Growth and the Import of Technology: Poland 1971-1980", *Cambridge Journal of Economics*, Vol. 2/1, March 1978.

"Britain's Slow Industrial Growth: Increasing Inefficiency versus Low Rate of Technical Change", in W. Beckerman (ed.), *Slow Growth in Britain: Causes and Consequences*, Oxford University Press, 1979.

"Industrialisation and the Rate of Growth: Eastern Europe 1955-1975", *Journal of Post-Keynesian Economics*, Vol. V(3), Spring 1983.

Alec NOVE is a fellow of the British Academy and Emeritus professor of Economics at the University of Glasgow. He was formerly Director of the Institute of Soviet and East European Studies of the University of Glasgow. His publications include:

The Soviet Economic System, George Allen & Unwin (publishers) Ltd., 1977.
Political Economy and Soviet Socialism, George Allen & Unwin, 1979 and *The Economics of Feasible Socialism*, George Allen & Unwin, 1983.

George HOLLIDAY is a Specialist in International Trade and Finance with the Congressional Research Service, Library of Congress, Washington, D.C. In that capacity, he has contributed to a number of Congressional and other publications on such topics as East-West trade, technology transfer, US export policy and international finance. His publications include:

US-Soviet Commercial Relations: The Interplay of Economics, Technology Transfer and Diplomacy (with John P. Hardt), House Committee on Foreign Affairs, Washington, D.C., 1973.

Technology Transfer to the USSR, 1928-1937 and 1966-1975: The Role of Western Technology in Soviet Economic Development, Westview Press, 1979.

"Western Technology Transfer to the Soviet Union: Problems of Assimilation and Impact on Soviet Exports", in *The Soviet Economy in the 1980s: Problems and Prospects*, US Congress, Joint Economic Committee, Washington, D.C., 1982.

Book I

ECONOMETRIC EVALUATION OF THE CONTRIBUTION OF WEST-EAST TECHNOLOGY TRANSFER TO THE EAST'S ECONOMIC GROWTH

by

Stanislaw Gomulka and Alec Nove

Part A

INTRODUCTION, THE MEASUREMENT PROBLEM AND PRELIMINARY EVIDENCE[1]

1. INTRODUCTION

The topic of this paper is important from the standpoints both of theory and policy. The policy implications are self evident. They are particularly clearly visible in the context of the recent conflict between American and West European policies with regard to East-West technology trade in general, the gas-pipe project in particular. This conflict is in part based on non-economic considerations, such as differing perceptions as to the internal system and global intentions of the Soviet Union, matters which fall outside the scope of our paper. However, one aspect of the transatlantic dispute is directly relevant to our theme: it concerns the degree of Soviet vulnerability to Western economic pressures, the extent of Western "leverage". This aspect too raises questions on matters other than technology transfer, e.g., the interest rate on credits or supply of grain. However, a large-scale acquisition of up-to-date Western technology is widely held to be of particular importance for the economic health of the USSR and Eastern Europe. No one doubts that, through the exchange of Western machinery and technology for oil, gas and gold, the USSR seeks to strengthen its economy. (Whether this should of itself be a potent argument against East-West trade, and whether we have any reason to suppose that the East benefits more than the West from such trade, are questions we will leave aside.) It seems legitimate to attempt quantification. How important is, then, the role of Western technology for the USSR and Eastern Europe? This is the key question we shall be concerned with. The answer could also cast light on the possible consequences of denying Western technology to the Soviet Bloc in the future, and thereby make a contribution to the policy debate that is still very much in progress. Thus if, as has been asserted, Western machines are in terms of net output per one rouble spent to obtain them, some ten times as productive as their Soviet counterparts, it can be held that the West makes, through machinery exports alone, an important contribution to Soviet growth, and that this represents a strong bargaining counter.

However, attempts to measure that contribution raise major problems of theory and empirical methodology. Firstly, new technology, whether imported or domestic, is only one of many factors in the growth of any country. The output is usually a joint product of all these factors. Yet here we will be discussing attempts to separate out and quantify the contribution of imported Western technology only. Is such a separation possible? Secondly, in the USSR as elsewhere, technological progress may depend on the investment necessary to ensure the internal diffusion of long-known technology, or on the less wasteful use of energy and materials by the old enterprises, or on a more intelligent choice between new investment alternatives. So the measurement of the separate contribution of technology, even if possible theoretically, is in any country very difficult empirically, the estimates depending to a distressing extent on the particular assumptions which are built into the model by the analyst.

In the USSR, imported Western machinery represents only about 5-6 per cent of the total of machinery and equipment currently installed. This percentage reached its peak in 1976-77, and has since been falling (Hanson, 1982). The rate of Soviet economic growth has also fallen in the years since 1976, and it may seem tempting to say that the latter was caused by the former. It is perhaps right to warn the reader already in these introductory notes that this would be an illegitimate instance of *post hoc ergo propter hoc,* as output and labour productivity grew much more rapidly in the period, say, 1955-1965, when imports of Western technology were far lower. Evidently, other factors were decisive.

There have been some valuable studies which bear on our theme by scholars who are participants in the OECD project; notably the work of Philip Hanson, Ronald Amann and Julian Cooper. These have, in the main, concentrated not on econometric macro-analysis, but rather on specific sectors, tracing the relative speed with which the technology has been incorporated in productive processes, comparing this with Western countries, identifying specific industries (such as the chemical industry) and investment projects to which the contribution of Western technology has been particularly significant. As will be clear from what follows, we greatly welcome such studies.

The object of this paper, however, is to relate and examine critically the attempts that have been made to *quantify* the impact of imported technology on Soviet and East European economies, by applying econometric and other quantitative methods, to *aggregate* economic statistics for the industry total and selected industrial branches. The studies reviewed are those by Donald Green and Herbert Levine, Martin Weitzman, Steve Rosefielde, Yasushi Toda and Stanislaw Gomulka. We do not deal with detailed micro-economic, project-based studies, such as would – for instance – tell us what kinds of Western equipment have been imported for use in the oil or the chemical industries and their impact there on output and efficiency. We again stress that such detailed studies are valuable, and many of the criticisms which we (and others) levy against the macro-economic calculations do not apply to them. On the other hand, the micro-economic studies have also evident limitations. While indicative of things larger, they remain local and specific. However, imported technology may benefit not only or mainly the users of the goods supplied by the investment project or the sector to which it is devoted. Such economy-wide indirect benefits are easy to miss in any micro study. An example would be a machine which produces a new product which then enables production elsewhere of some other products to be undertaken cheaply, or technical/managerial expertise which is gained in one project, but diffused later throughout the industry. In principle, these kinds of effects should be more easily "caught" by macro-economic analysis, which seeks to link total technology imports, or more often total machinery imports, with the growth performance of the whole economy or its major sectors. The studies we review use different methods and unfortunately report different results. Still the studies are useful, although major methodological difficulties and data deficiencies stand in the way of obtaining estimates that can be taken as reliable.

Technology transfer takes, of course, many forms, and the actual movement of machinery and equipment from West to East is but one of them. However, the impact of machinery transfer seems the easiest to measure, and therefore most of the econometric work which we review in this paper concerns this particular form. But we shall stress that other channels must not be disregarded, especially in the case of the USSR. One such channel is the transfer of know-how, of patents and the like, which enables the Soviet R & D sector to invent their own processes or products better and quicker than would otherwise have been the case. The other is the acquisition of operational or production-engineering knowledge (lay-out of factory, arrangement of equipment on the production line, etc.). To both these ends, the Soviets obtain scientific and technical publications on a large scale, and take opportunities to study Western scientific-technical achievements – though this may do no more than counterbalance the loss to them arising from the extent to which most of their scientists, engineers and managers are cut off from direct contact with the West.

Our review of the studies makes us emphasize rather forcefully the possibility that imports of Western capital *goods* may not, in fact, be the most important way by which the Soviets acquire Western technology and know-how. The huge Soviet R & D establishment itself acts as a massive pump that identifies, decodes and translates Western advances in science and technology into Soviet industrial applications. The relatively minor Soviet imports of investment, and indeed consumer, goods may well have a relatively *small direct contribution* to Soviet economic growth, as most of the studies we review suggest. However, these imports, along with the other agents of the transfer of know-how we mentioned above, may still play a vital stimulating role for the Soviet R & D in its identification, translation and development work. There are grounds to think that the contribution to growth of such indirect effects is the more important one and possibly large as well.

2. TECHNOLOGICAL CHANGE AND OUTPUT GROWTH: THE MEASUREMENT PROBLEM IN THE SHORT RUN AND LONG RUN

In view of the interaction between technological change and capital accumulation, to which we allude in the Introduction above, it may be useful to define more explicitly the measurement problem in growth accounting in general and the meaning of any measure of the separate contribution of technological change to economic growth in particular. This lengthy section is thus still introductory in nature, and somewhat theoretical in orientation, preparing the ground for the arguments of the main body of the text.

All production inputs in any economy may be classified into the following four categories: labour force L, physical capital (including land) K, human capital H, and technology T, the latter meaning the state of knowledge relevant to production. Human capital refers to the skills embodied in the labour force, reflecting its familiarity with, and the capability of, using the available technology, the latter in turn being the knowledge embodied in physical capital. Flows of services of these inputs are used in *co-operation* in a large number of separate production processes, the latter grouped into different activities of individual enterprises, to produce the following flows of output: consumption C (which may include additions to knowledge not useful for the purpose of production), additions to K, additions to H, and additions to T. Aggregating over these activities, we may thus write that $Y = F(L, K, H, T)$, where Y represents the flow of aggregate output and L, K, H and T denote now the flows of services of the respective inputs. The labour and capital services are measured in man-hours and dollar-hours. We do not have a convenient measure for the services of H and T, but we associate all qualitative changes in the economy with a rise in T or H, or both.

Since the inputs are used in *conjunction,* it is not useful to ask what is, at any particular moment of time, the contribution of any one input to the *total* flow of output.

If all inputs of any of the input categories were zero, the output would be zero or near to it. In this sense each of the aggregate inputs is essential, and all are responsible for all the output.

Therefore, we usually consider small *changes* in outputs, such as those occurring in a one-year period, and ask what could be the change in output if any particular input changed as it did while other inputs were constant. When the changes in aggregate inputs are small, it is legitimate to presume that the resultant output effects are independent each of the other. It is this *independence property* that economists assume to exist and it is this property which enables them to separate out the individual contributions of the small changes in inputs to the total change in output.

To give an example, suppose that five men with picks and shovels are replaced by one man and one bulldozer (which may be domestically produced or imported). Output rises. So does labour productivity, the labour used being of higher skill. In both processes the output is a joint product of men (with their different skills) and their tools (with the relevant different technologies embodied). In each process there are no such things as separate contribution of capital, labour and technology. However, the combinations of inputs, fixed in each process, can be varied if one considers using the two processes with different intensities. Suppose we take the number of five-men teams large enough so that the value of their picks and shovels is the same as the value of one bulldozer. The difference in output may then be attributed only to the difference in labour and technology. When a large number of different processes is considered, it is possible to combine them in such proportions that given labour and capital (or changes in labour and capital) only output and technology would vary, or given capital and technology, only output and labour would vary. Given these variations, one can then attribute changes in output to the corresponding (separate) changes in technology, labour or capital. This is the basis on which the production function estimates rest.

It should be stressed, however, that the independence property, as defined above, is not maintained when one considers a process of economic growth over a substantial period of time. For, a change in any one input at the beginning of the period would change the output flow over the period and, given the distribution of output between consumption and additions to inputs, would also be changing the supply of the other inputs, thus giving rise to further, indirect, changes in the output flow.

To illustrate this difference between short run and long term more sharply, let us simplify further by writing that $H = H(t)$, $T = T(t)$ and, therefore, $Y = F(L, K, t)$ where t represents time. Assuming constant returns to scale in terms of L and K, the equation which this production function implies is as follows:

$$g_Y = \pi g_K + (1-\pi) g_L + \lambda \tag{1}$$

where g is the growth rate of the variable indicated by the subscript, π is the elasticity of output with respect to capital and λ, equal to $\frac{1}{Y}\frac{\delta F}{\delta t}$, is the contribution of the qualitative changes to the output growth. This direct contribution is of course instantaneous or short term. If qualitative changes have been absent in the past, but will be present and maintained at the rate λ in the future, then not only g_Y would rise, but also g_K, producing an additional, indirect contribution to g_Y. What is its size? The answer depends on the decision concerning the distribution of the direct contribution mentioned above between investment in physical capital and other uses. Suppose we keep the share of capital investment in output unchanged. Then a rise in g_Y by λ increases immediately the growth rate of investment by λ and, over time, it also increases g_K by λ, producing a feedback effect on the growth rate of output of a size $\pi\lambda$. This additional increase in g_Y raises the growth rate of investment, and hence g_K, further by $\pi\lambda$, having an additional feedback effect on g_Y of a size $\pi^2\lambda$. The appearance of qualitative changes at a rate λ is seen to generate what may be called a *growth propagation process*, the contribution of which to g_Y is equal to $\lambda + \pi\lambda + \pi^2\lambda + ... = \frac{1}{1-\pi}\lambda$. The long term dependence of g_K on λ has thus the effect of multiplying the direct contribution λ by the multiplier $\frac{1}{1-\pi}$. This full contribution of qualitative change to growth is seen separately if (1) is rearranged to read:

$$g_Y = \frac{\pi}{1-\pi}(g_K - g_Y) + g_L + \frac{1}{1-\pi}\lambda \qquad (2)$$

Let v represent the ratio K/Y. Note that $g_K - g_Y = \dot{g}_v$. Hence (2) may be written in the form $g_Y - g_L = qg_v + \alpha$, where

$$q = \frac{\pi}{1-\pi} \text{ and } \alpha = \frac{\lambda}{1-\pi} \qquad (3)$$

The difference between λ and α is very important and should be well understood. In particular, it should be noted that α, although referring to a period of time, is not necessarily the actual contribution of technological change during that period. Since both λ and π, defining α, are current, α is only an approximation, or a forecast, of the true contribution of technological change to growth over any particular period. This forecast is based on short-term information, so it will prove correct in the unlikely event that the information remains unchanged over time. Moreover, our "growth propagation" argument assumes that both capital and output grow at a common rate and that both the savings ratio and the employment growth rate are constant. These are the properties of an economy that moves along a balanced growth path. It follows that while λ is an exact measure of the direct contribution of technological change to output growth, α is a *balanced growth* estimate of the total actual contribution over a period of time.

Since net capital investment $\dot{K} = s^Y$, and since $K = v^Y$, it follows that $\dot{K} = \dot{v}^Y + v^{\dot{Y}}$ and therefore (the dot over a variable denotes time differentiation):

$$s = vg_Y + \dot{v} \qquad (4)$$

Relations (2) and (4) are correct whatever the change in v over time. The capital/output ratios actually observed appear to be much more stable in the long term than labour/output ratios. For those situations when it is legitimate to consider an economy's or sectoral capital/output ratio to be in fact constant, relations (2) and (4) assume a particularly simple form:

$$g_Y = g_L + \alpha \text{ and } s = vg_Y \qquad (5)$$

We do not have reliable methods of estimating and π and λ. However, in those cases in which the capital/output ratios are, in the long term, approximately constant, the long-term contribution of qualitative changes to output growth is seen from (5) to be (approximately) equal to the growth rate of labour productivity, and this we can measure more reliably. We also infer from (5) that the wide international diversity recorded in the growth rate of labour productivity may be approximately equal to, as well as implied by, the differences in the rate of qualitative changes.

When the labour participation rate is constant, so that the nation's population and employment in manhours change at the same pace, and when the savings rate is also constant, so that consumption and output grow at a common rate, then α is seen from (5) to equal the rate of growth of consumption per person. The labour participation rate and the savings rate may and do change with time, but when they are relatively stable, as they in fact tend to be over substantial periods of time, then it is correct, in the sense given above, to attribute approximately all the long-term rise in the average consumption and income per person to the past qualitative changes that have accumulated over time in any one country. Capital accumulation still remains essential as a necessary medium without which qualitative changes cannot be fully effective in the short run and cannot take place in the long run.

This intimate interaction between qualitative changes and capital accumulation makes a growth model, and not merely a production function, the appropriate instrument for measuring the contribution of qualitative changes to growth over a period of time, and for contrasting the contribution of a change in the savings ratio, rather than in the growth rate of capital stock, with the impact of a change in α. To put it another way, in the short term the marginal productivities of all inputs are important and need to be known to assess their respective contribution to growth. However, over a long period of time, such as the post-war period, only the rates at which the productivities of the non-producable (or primary) inputs change are crucial. Moreover, the (aggregate) labour productivity growth rate is usually a good approximate measure of these changes. This point is too often missed or ill-understood by economic writers, although of course well known in the theoretical literature on economic growth; it is particularly well discussed by Dan Usher (1980) in a chapter entitled significantly and appropriately "No technical change, no growth". In that chapter the reader will find a careful and extensive discussion not only of the production function estimates of the contribution of technological change to economic growth by Solow (1956), Denison (1962), and Jorgenson and Griliches (1967), but also the discussion of the changing qualities of the inputs and outputs, and of the use of the Divisia index for aggregation purposes, as sources of bias in such estimates.

Correction for the changing (static) x–inefficiency

In writing $Y = F(L, K, t)$ we meant to capture the grand production relationship between "potential" inputs and outputs in the neighbourhood of the actually observed ones, and we assumed that the latter are aggregates over efficient techniques, implying "full" utilisation of all the inputs. This assumption is unrealistic not only in the short run, say during a depression, but also in the long run. Let u_K and u_L be the utilisation rates of K and L respectively, calculated at some "standard" intensity of work. Hence $Y = F(u_K, K, u_L L; t)$. Now instead of (1) we have $g_Y - g_L = q(gv + g_{u,K}) + g_{u,L} + \alpha$, where $g_{u,K}$ and $g_{u,L}$ are the growth rates of u_K and u_L respectively. When qg_v can be assumed to be a minor term, we have

$$g_Y = g_L + \alpha + g_u \quad \text{and} \quad s = vg_Y \qquad (5a)$$

$$\text{where } g_u = qg_{u,K} + g_{u,L}.$$

These are two key aggregate growth relations, useful as a starting point in any analysis of the contribution of technological change to long-term growth.

3. THE PRELIMINARY STATISTICS

The pace of Soviet industrialisation since 1928 has been somewhat slower than in Japan, but of comparable speed or faster than that experienced in Western Europe and the United States. The result is that, in 1976, "in ruble prices, Soviet GNP was 50 per cent of United States GNP; in dollar prices, the USSR produced final goods and services equal to 74 per cent of the United States national product"[2]. According to the same source, the outputs of the two key industrial sectors, one producing machinery and equipment for investment purposes and the other defence goods, are thought to have outpaced the corresponding outputs of the United States industries[3]. An extraordinary emphasis has also been given to technical education, basic science and industrial research, and technological progress. According to the Nolting-Feshbach study the number of Soviet R & D scientists and engineers in 1978 was "nearly 60 per cent greater than the United States"[4]. The non-personnel expenditure on R & D is now probably of comparable size in both countries. If we take the trend rate of industrial labour productivity growth for a measure of the technological progress (the one denoted by α in the previous section), we find that, at about 5.5 per cent in the years 1928-1975, it was comparable to that of Japan and Western Europe in that period, and significantly higher than in the United States and the United Kingdom.

On the other hand, a large body of evidence has been accumulated to indicate a high degree of resource misallocation, both in conventional production and R & D, and a large but rather slow and often wasteful investment activity. The important evidence is the one indicating that despite its large size, the contribution of Soviet and East European R & D activity to the world flow of new inventions is apparently minor. Since it is this type of evidence which gives grounds for the view of Western technology playing a vital role in East's technological progress and productivity growth, it may be useful to give it in some detail. The important reference is the study by Jiri Slama (1982). His sample of 28 countries includes 20 OECD countries (all except Austria, Iceland, New Zealand and Turkey), seven centrally planned economies, and Yugoslavia. The seven include the USSR and six countries of Eastern Europe. According to this study, the Seven's share of the relevant total for the whole sample is 53 per cent of the R & D personnel, but only 36 per cent of the domestically-registered patents. Perhaps even more significant is the fact that, in terms of the number of patents registered abroad, the inventive output of the Seven's R & D sector is only about one-thirtieth of the West's output, indicating that most of the domestically-registered patents of the Seven are small improvements of little or no international interest. Another interesting fact is that in the 1970s the Seven imported about ten times more licences, in terms of dollars paid, than they exported, the exports representing merely about 1 per cent of the estimated total of world exports. Similarly insignificant is the Soviet and East European share of technology-intensive, as well as all manufacturing, exports in the Western world market. Martin Spechler notes that "Soviet reluctance to allow their best technicians and engineers to spend extended time in the free world makes it rather unattractive to buy sophisticated and complex technology from them, as such systems require knowledgeable installation and post-sale service, not to speak of ready spare parts" (1982, p. 21). This factor does play a role, but the poor quality of goods supplied for the Soviet and East European domestic markets suggests that lagging technology and poor management are by far the more important factors. The USSR clearly has been, and in most areas continues to be, a catching-up country. The economic growth of any such country is based largely on the gradual absorption or adaptation of the outside technology. However, that outside technology may be channelled in a variety of ways, the import of machinery (and licences) being just one of them. The controversy concerning the impact of West-East transfer of technology is, or should be, not so much about the total impact, as about the contribution of the machinery imports alone. We reproduce some of the data about these imports in the Appendix 2.

Part B

THE WORK DONE AND THE RESULTS REPORTED: A SURVEY

1. *Green-Levine:* SOVMOD III, as reported in "Macroeconometric Evidence of the Value of Machinery Imports to the Soviet Union", Stanford Research Institute, 1977, and in "Soviet Machinery Imports", *Survey,* Spring 1978, Vol. 23, No. 2, pp. 112-126.

The authors seek to estimate the magnitude of the impact of imported machinery from the West on Soviet industrial output in the period 1960-1974, as well as to evaluate probable gains of this kind in the period 1975-1980. The domestic value of the stock of machinery of Western origin is estimated for the years 1960-1980 for industry total and for three branches: petroleum products, chemicals and petrochemicals, and machine building and metal working. These data imply that the ratio of accumulated investment of Western origin to total capital stock, K_i/K, was in the years 1960-1974 between about 1.5 and 2 per cent for industry total, and between about 3 and 6 per cent for the three branches. This ratio was, generally, increasing during the period, but so slowly that the cumulative change from 1960-1975 is almost insignificant.

Capital K_i, which represents only machinery, is treated as qualitatively different from the capital of non-Western origin, $K-K_i$, which includes both machinery and structures. The authors take as their basic assumption that Soviet output levels are determined by three separable factors. These are: capital imported [from the West, but in the case of the MBMW Branch (Machine Building and Metal Working) also from the CMEA], any other capital (mainly domestic), and the labour employed. The authors examine not simply the situation in a single year, but the year-to-year growth in inputs and outputs. To the extent that the *productivity* of different input combinations vary, when expressed in terms of rubles of output per unit of composite input, the quantities of the input and the proportions in which they are combined affect the rate of growth of output. The question is by how much. The authors' immediate objective is therefore to estimate, by means of econometric techniques, the marginal productiveness of the three inputs mentioned. It then becomes possible

 i) to quantify the role of the three inputs in past growth and
 ii) to estimate the effects on future growth of Soviet imports of Western machinery.

The authors assume that the production function that links the three inputs is of the Cobb-Douglas type:

$$Y = A \, K_d^{\alpha} \, K_i^{\beta} \, L^{\gamma} \qquad (6)$$

where Y is output (value added), $K_d = K - K_i$, and L represents the labour inputs in manhours. The elasticities α, β and γ determine the marginal productivities we mentioned above. The elasticities need not sum up to unity, so non-constant returns to scale are admitted. However, we notice at once that the above specification of the production function *assumes* imported capital to be an essential input; output would be nil in the absence of these imports (for a discussion of this point see Part C). Moreover, (6) has no separate term for technical change; if the quantities of the three inputs were to be constant, increasing qualities of these inputs would have, then, no effect on the volume of output. Both assumptions will have important implications to which we shall return in Part C.

Expressed in growth terms, Eq. (6) takes the form:

$$g_Y = \alpha g_{K_d} + \beta g_{K_i} + \gamma g_L \qquad (7)$$

where g_Y represents the growth rate of Y, g_{K_d} the growth rate of K_d, etc. In this equation βg_{K_i} represents the contribution of Western new capital to the growth of output. By definition of the marginal product, $MP_{K_d} = \alpha Y/K_d$ and $MP_{K_i} = \beta Y/K_i$, where MP stands for marginal product. Hence the ratio of the two marginal products, MP_{K_i}/MP_{K_d}, is equal to $\frac{\beta}{\alpha} \frac{K_d}{K_i}$.

Green and Levine estimate first the elasticities α, β, γ, and then they find this ratio of marginal products.

Their estimate of the ratio, and this is the main result of this work, implies that, at the margin, the ruble's worth of imported machinery was, in the period 1961-1974, as productive as 8.1 rubles of non-Western capital in chemicals, petrochemicals and petroleum products, as 21.9 rubles in machine building and metal working, and as 14.0 rubles in industry total. The estimate for industry total implies that a 10 per cent growth of machinery imports from the West (mean annual for 1961-1974 was 10.5 per cent) adds 1.1 per cent to the annual growth rate of the Soviet industrial output[5]. Moreover, following our discussion of the growth propagation effect in the previous Part, suppose this and any further gain in output growth enables the USSR to increase the growth rate of non-Western capital by the same magnitude, as one may expect to happen if investment changes proportionately to output. This would then make possible a further increase in Soviet industrial growth by

$$(\alpha + \alpha^2 + \alpha^3 + \ldots) 1.1\% = \frac{\alpha}{1-\alpha} 1.1\% = 0.8\%,$$

so that the total (direct and indirect) gain in output growth would be 1.9 per cent, or about a quarter of the Soviet industrial growth in that period.[6]

In a situation of balanced growth, when the growth rates of Y, K_i and K_d are all the same, and this is in fact what one might expect to be approximately the case in the long term, we would have that (this follows from Eq. 7):

$$g_Y = \frac{\gamma}{1-\alpha-\beta} g_L = g^*_Y, \text{ say:} \qquad (8)$$

If machinery imports from the West were to stop from now, Soviet industrial growth rate g_Y would converge, overtime, to a value of $\frac{\gamma}{1-\alpha} g_L$. This is implied by (7) when $g_{K_i} = 0$ and $g_Y = g_{K_d}$. Therefore, the Green-Levine specification of the Soviet production function implies that, in the absence of any change in K_i, the Soviet industrial balanced growth rate would drop by $\frac{\beta}{1-\alpha} g^*_Y$, which is the difference between g^*_Y, as given by (8), and $\frac{\gamma}{1-\alpha} g_L$. The ratio $\frac{\beta}{1-\alpha}$ gives thus the drop as a proportion of the g^*_Y rate. The Green-Levine estimates of α and β imply that this proportion would be equal to 0.26 in machine building and metal working, 0.41 in petroleum products, 0.11 in chemicals and petrochemicals, and 0.20 in total industry[7]. These four numbers represent their balanced growth estimates of the losses the USSR would suffer if further Soviet machinery imports from the West were to stop, the losses expressed as fractions of the output growth rates in the listed sectors, the growth rates being achieved in the presence of the imports.

These ratios may be looked at as either potential losses or actual contributions. They are seen to imply very substantial contributions indeed from Western machinery imports. The respective contributions in the medium-term are found lower, yet still significant. An experiment with SOVMOD III gives for 1973, in terms of the 1968 levels, the following contributions: 1.1 per cent of the GNP, 2.5 per cent of total industrial output, 7.4 per cent of the machine building output, 1.8 per cent of total consumption. Another experiment with the same model implies that 1 ruble of machinery imports in the years 1973-1978 increased Soviet industrial production in that period by 8 rubles. This is possibly the most representative figure for the Green-Levine study.

2. *Weitzman:* "Technology Transfer to the USSR: an Econometric Analysis".
   Toda: "Technology Transfer to the USSR: the Marginal Productivity Differential and the Elasticity of Intercapital Substitution in Soviet Industry".

Both studies cover the period 1960-1975 and are based on the same data as the Green-Levine study. Their primary aim is to find out whether the Green-Levine estimates would survive if other plausible specifications of the Societ production function were tried.

One such specification tested by Weitzman allows for technical change, but assumes constant returns to scale. This is as follows:

$$Y = A e^{\lambda t} K_d^\alpha K_i^\beta L^{1-\alpha-\beta} \tag{9}$$

where λ, the rate of technical change, is allowed to be either a constant or changing with time in a linear fashion: $\lambda = \lambda_0 + \ell t$. Weitzman reports that in all eight cases considered (4 branches times 2 specifications of λ), with the single exception of petroleum, the estimates of β are either negative or insignificantly different from zero, implying negative or minor marginal productivity of imported Western capital.

The elasticity of substitution between K_d and K_i, let it be represented by σ_{KK}, is in (9) unity, and is the same as that between capital and labour, σ_{KL}; consequently all inputs are essential for any output, as in the Green-Levine study. The other four specifications tested by Weitzman incorporate different assumptions concerning these two elasticities and λ. In variance with the Green-Levine study, it is assumed that returns to scale are constant and, more significantly, that $\sigma_{KK} = \infty$. This latter assumption seems particularly restrictive, since we would expect lower than perfect substitutability between imported and domestic machinery if they are qualitatively different. Weitzman takes the view that "the elasticity of substitution between domestic and imported capital ought to be far greater than the elasticity of substitution between capital and labour", and he estimates the latter to be less than unity. The two types of capital are perhaps less dissimilar than capital and labour, so we too would expect $\sigma_{KK} > \sigma_{KL}$, but his four specifications are clearly an extreme representation of this view, and this might be considered to be a problem area, perhaps even a major flaw, of the study. Domestic capital and imported capital might be assumed to be highly, even perfectly, substitutable only if they were technologically similar. But it is precisely this similarity, or dissimilarity, that Weitzman wants to establish by his test. The assumption of $\sigma_{KK} = \infty$ amounts to taking a stand before a test on a matter that is the subject of the test.

It is therefore interesting to have the results obtained by Yasushi Toda, who relaxed Weitzman's assumption concerning σ_{KK}. In a paper published in the same June 1979 issue of the *Journal of Comparative Economics* he reports his estimate of σ_{KK} to be close to unity. This vindicates the Green-Levine assumption. But his statistical tests support also Weitzman's thesis that "the marginal productivities are indistinguishable between domestic and imported capital". The absence of the technical progress term appears thus to be the main reason underlying the Green-Levine finding, that a substantial difference exists.

Returning to Weitzman's study, his four specifications are as follows:

i) λ any constant
ii) $\lambda = \lambda_0 + \ell t$ $\Big\}$ $\sigma_{KL} = 1$;

iii) λ any constant
iv) $\lambda = \lambda_0 + \ell t$ $\Big\}$ σ_{KL} any constant.

It may be seen that (*i*) is a special case of (*ii*) and both (*ii*) and (*iii*) are special cases of (*iv*). In all these specifications the aggregate capital K is a weighted sum of K_d and K_i, that is

$$K = K_d + wK_i \qquad (10)$$

where the weight w is assumed to be a constant, independent of the ratio K_d/K_i. It is this assumption which implies perfect substitutability between K_d and K_i. It is not unreasonable to expect that in some cases any job that a Western machine does can be also done by a Soviet machine or machines. Suppose that in these cases 1 ruble of Western machinery can be substituted by w rubles of Soviet machinery. One would expect, however, that this substitution ratio varies significantly from project to project. Presumably the Soviets attempt to rank Western machinery in terms of w and import first of all those with the highest rank. The average rank would then vary with the size of total machinery imports. Weitzman's assumption excludes this, as if all machines had the same rank, or the same efficiency advantage. If the capital imported from the West were qualitatively identical to that of non-Western origin, then w would be unity. Weitzman seeks to find the value of w which fits best Soviet growth data, given the above four specifications. To this end he fixes w at different values and finds the quality of the fit in each case. It turns out that the growth data are "explained" almost equally well at widely different values of that parameter. The optimal values (at which the error sum of squares is at a minimum) turn out to vary from as low as −12 [industry total, specification (i)] to as high as 1 000 [chemicals and petrochemicals, specification (ii)], and the standard errors of these estimates to be so great that the true value of w could be almost anywhere in this range.

The conclusion to be drawn from this study is that while it is possible that the productivity of imported Western capital is higher than that of domestic Soviet capital, there are at the moment no sound statistical grounds to maintain that this is, in fact, the case. Therefore, contrary to Green and Levine, Weitzman's study implies that the contribution of imported Western machinery could well be significant, although equally there are no sound grounds to maintain that this is, in fact, the case.

3. *Rosefielde:* "East-West Trade and Postwar Soviet Economic Growth: A Sectoral Production Function Approach"

Stephen Rosefielde comes to conclusions similar to those of Weitzman, albeit for different reasons. Using a number of methods, the author obtains several estimates of the sectoral rates of technological change in the years 1950-1973. He seeks to establish whether or not these rates are higher in sectors/periods-of-time where/when the machinery imports are also higher. The answer to this question is found to be negative: the rate of technological change appears to be unrelated with the share of imported Western capital in the total stock of fixed assets. Rosefielde accepts, however, that "for several decades, the rate of technical progress achieved by the Soviets exceeded that of the US, resulting in an obvious closing of the relative technological gap" (p. 105), but he believes that Soviet technical progress "must largely be attributed to domestic Soviet science and engineering" (p. 111). And he concludes that his "analysis suggests that the magnitude of the Soviet achievement in borrowing efficiently and cheaply from the West has been universally and significantly exaggerated" (p. 134). This conclusion appears to us to be a much stronger view than a prudent analyst would take on the basis of his study's evidence.

Nevertheless, one does not exaggerate much by saying, as Martin Weitzman, commenting in the same study on Rosefielde's results does, that "the evidence so far is that we cannot determine any influence of technology transfer (read: imports of Western machinery) on Soviet economic growth using existing data. It's too bad, but nothing consistent seems to emerge from the numbers" (p. 140).

Yet, it is quite possible that while the narrowly defined direct impact of machinery imports on technological change and growth is insignificant, this and other forms of diffusion of Western technology may act as powerful stimulators which significantly increase the productiveness of the Soviet science and engineering. Rosefielde's tests were not designed to test this possibility.

4. *Gomulka:* "Inventive Activity, Diffusion and the Stages of Economic Growth", Aarhus Institute of Economics, 1971; "Import-led Growth: Theory and Estimation", 1976; "Growth and the Import of Technology: Poland 1971-1980", *Cambridge Journal of Economics,* March 1978.

The first of these three works is largely a theoretical study of macro-effects of international technological diffusion. One of its main propositions is that differences in international technological rates of diffusion, rather than in original domestic R & D activity, are the major force underlying the differences among nations in productivity growth rates in the 20th century.

The evidence that diffusion is, for most of the countries, that found themselves behind the technological frontier, the dominant determinant of their technological advance is both direct and indirect. The direct evidence concerns particular countries, such as Japan in the 1960s or the Middle East now, and suggests that nearly all new products and processes adopted there are either of foreign origin or represent relatively simple extensions of imported technology.

The indirect evidence originates from comparing the expansion of the technological sector in various parts of the world. For instance, this sector in Europe and the USSR has been expanding about as rapidly as that in the United States. Yet, in the postwar period the rate of labour productivity growth, which in the situation of approximately stable capital/output ratio may be used as a proxy for the (long-term or dynamic) rate of technological change, has been nearly twice as high in Western Europe and the USSR than in the United States. The implications of this fact are immediate and important. Denoting $\Delta T = \Delta T_d + \Delta T_f$, where ΔT_d is the domestic output of new technology and ΔT_f is the inflow of foreign and foreign-induced technology, we have

$$\frac{\Delta T}{T} = \eta \frac{\Delta T_d}{T_d} + (1-\eta) \frac{\Delta T_f}{T_f} \tag{11}$$

where $\eta = T_d/T$. That is to say, the country's rate of technological change is a weighted sum of the rate generated by the domestic research activity and the rate generated by assimilation of foreign innovation. The growth rate $\Delta T_d/T_d$ is related to the trend growth rates of domestic R & D inputs. Over the last half century these have been, compared with the United States, somewhat higher in the USSR and lower in Western Europe. Assuming about the same productivities of these inputs, an assumption which probably favours the USSR, one should expect $\Delta T_d/T_d$ to be about the same in all the three areas. Therefore, substituting $(\Delta T/T)$ United States for $\Delta T_d/T_d$ in (11), one would have

$$(\frac{\Delta T}{T})_{USSR} = \eta (\frac{\Delta T}{T})_{USA} + (1-\eta)(\frac{\Delta T_f}{T_f})_{USSR} \tag{12}$$

Since $0 < \eta < 1$, the only way to explain why $(\frac{\Delta T}{T})_{USSR} > (\frac{\Delta T}{T})_{USA}$, over the years 1928-1978, is to admit that

$$(\frac{\Delta T_f}{T_f})_{USSR} > (\frac{\Delta T}{T})_{USSR} \tag{13}$$

in that period. But (13) says that the stock of foreign-originated technology, transferred to the USSR or adapted locally, has been expanding faster than the stock of Soviet original technology. Moreover, as long as (13) holds, the ratio T_d/T is declining, pointing to an increasing weight of Western or Western-derived technology in the total stock of Soviet technology. (The same point applies to individual countries of Western Europe in the postwar period.)

The technological diffusion from the West is, according to this argument, capable of producing a rate of technological progress in the USSR over and above that in the

United States only as long as the technological gap between the USSR and the West remains sufficiently large. As long as this gap falls, converging to a certain equilibrium level (the magnitude of which need not be zero, as it is likely to depend strongly on various systemic and cultural factors), the (growth) rate of technological diffusion falls too, bringing about an equalisation in productivity growth rates between the United States and the USSR. In equations (11) to (13), $\Delta T_f/T_f$ falls to the level of $\Delta T_d/T_d$, so that the ratio T_f/T_d stabilizes; technological transfer would be larger than ever, but it would cease to be capable of contributing further to closing the technological and productivity gaps. The formal theory of technical change is however not very helpful in ranking the various forms of diffusion in terms of their contribution, direct and induced, to the total flow of a country's technological innovation.

The primary purpose of the empirical studies [Gomulka-Sylwestrowicz (1976) and Gomulka (1978)] was to evaluate the contribution of machinery imports from the West to, among others, the USSR and Poland. The method employed is not based on production function analysis of the Green-Levine-Weitzman-Rosefielde-Toda type, as small data sample and other problems are thought to make such an analysis unreliable. In fact the method is algebraic rather than being a purely econometric test. It seeks to calculate the productivity and output effects of machinery imports that would have been obtained if

i) the output per unit of labour is higher on imported machines than on domestic ones, but output per unit of capital is the same on both types of machines, the assumption thought to be realistic in view of the relatively high international variation in labour productivities and high similarities in capital-output ratios at the level of total industry in both West and East;
ii) imported machines going to the sector producing investment goods produce themselves machines of the same (labour and capital) productivity characteristics as their own; and
iii) the proportion of machines imported from the West going to the (investment) sector itself producing machinery equals the share of gross investment in gross national product.

Under these assumptions, and they may overstate the actual contribution of imports by a significant (unknown) margin, the diffusion effect was found to be insignificant for the USSR, but rather important for Poland, especially in the 1970s. This is because Poland imported in the 1970s, relatively speaking, about five times more than did the USSR.

The "rule of thumb" suggested by the above calculations for the industrial sector is as follows: an increase in the share of machinery imports from the West in all machinery investement by 1 per cent, if *sustained over a longer period of time*, results in an increase of the growth rates of labour productivity and output by about 0.1 per cent. It may be recalled that the Soviet Union's imports from the West amount to about 3 to 6 per cent of all her machinery investment, while the corresponding figure for East Europe is about 10 to 30 per cent. These figures imply that the contribution of the Western machinery imports to industrial growth is something like 0.3 to 0.6 per cent in the Soviet Union and between 1 and 3 per cent in the case of East European countries[8]. These are long-term contributions. The relevant Green-Levine estimate is about 2 per cent for the USSR. The conclusions which Gomulka's studies imply is that the transfer of technology from the West has been and continues to be very important for the USSR as a source of productivity growth, but that machinery imports are apparently a relatively unimportant channel of that transfer. By implication, all the other forms of diffusion, as well as original and (especially?) derived innovation by the Soviet R & D sector, would appear to be the more powerful agents of Soviet technological change. The conclusion is, of course, somewhat modified for the East European countries. The more general rule suggested by this work is that machinery imports are actually or potentially an important agent of technological transfer for any country with a relatively small R & D sector, while the so-called disembodied diffusion and induced innovation are relatively more important for countries with a large R & D sector, such as Western Europe, Japan, USSR and possibly China.

5. SUMMARY

The results reported by the studies above may be summarised as follows:

According to Green and Levine, imports of Western machinery are comparatively small in volume, but are, nevertheless, capable of increasing Soviet industrial growth by about a fifth in the long term, and must be regarded as an important source of Soviet growth in the short term too. Weitzman does not maintain this conclusion to be necessarily wrong. However, his econometric tests led him to take the cautious and somewhat agnostic view: one cannot say with confidence, given the data we have at the moment, what, in fact, is the value of Western machinery to the Soviet economy; it could very well be low. The Rosefielde study yields conclusions close to those of Weitzman. These studies are concerned exclusively with the evaluation of the technology transfer to the Soviet Union through machinery imports from the West. They do not discuss other forms of technology transfer nor do they cover other countries. They also share the method of analysis, which is based on estimating production functions. Gomulka's work differs from these studies in three respects. It covers some 40 countries, in particular the USSR, Poland and Japan; its theoretical part seeks to offer an integrated treatment of produced and diffused technical change within the standard growth theory; its empirical part employs an algebraic method, based on a suitably generalised vintage-capital model, rather than an econometric method of aggregate production function. This work suggests that the (direct and indirect) technology transfer from the West has been and continues to be of key importance to the Soviet and East European economies. However, the role of machinery imports from the West appears to be relatively minor for the USSR, but rather important for the East European countries. Other forms of technological and scientific diffusion to the USSR, which the West is largely unable to control, would therefore appear to play a relatively greater, possibly much greater role in contributing to and stimulating the Soviet technological advance[9]. His studies also imply that the rate of technological advance in the USSR will soon be (or perhaps is already) declining, and if machinery imports from the West are undertaken on a massive scale, and/or if a much more competitive economic system within the USSR is adopted, this decline might be averted, but it would be averted for a limited period only. This point is not intended to belittle the significance of such potential postponement in the otherwise inevitable decline of the rate of Soviet technological change. Its purpose is to bring home the point, which is somewhat paradoxical and not always well understood, that when actual technology is sufficiently close to the world's best technology, *increasing* Soviet (as well as West European or Japanese for that matter) domestic R & D effort and *increasing* technological imports from the West are consistent with *declining* rate of Soviet (as well as West European or Japanese) technological change. The objective of the Soviet increased innovational efforts is, or should be, to maintain the falling rate of technological change above the US level for some time yet, until the technological gap between the USSR and the United States is virtually eliminated, rather than to increase the rate or keep it at the past level. Even achieving this limited objective may not be possible under the present economic system, a well known view to which we shall return in Parts E and G of the paper.

Part C

CRITICAL COMPARATIVE EXAMINATION OF THE MEASUREMENT METHODS

We have already described, admittedly very briefly, the methods employed by the studies discussed above. It seems paradoxical that Green and Levine obtain high and statistically significant estimates of the marginal productivity of imported machinery while Weitzman, using the same basic method and data, does not. The reason for this appears to be the particular specification of the production function adopted by Green and Levine [see Eq. (6)]. This specification implies that Western machinery is, like labour or domestic capital, an essential input; Soviet industrial production would come to a *complete* halt in its *total* absence. This assumption is certainly widely unrealistic, though the direction of the bias it produces in the estimate of the marginal productivity of imported machinery is not clear. Conceivably, it may increase the estimate. Also unrealistic is the absence, in the Green-Levine specification, of a separate term for technical progress. Consequently, the growth effect of that progress has been captured by physical inputs. This itself tends to increase the estimates of α, β and λ, and therefore also the contribution of machinery imports to growth. Moreover, the rate of growth of the joint (capital and labour) Soviet productivity in Soviet industry, denoted λ in equation (1), was apparently relatively low in the years 1963-68[10]. Since in those years Soviet machinery imports happened to be also relatively low, it is possible that, in the Green-Levine exercise, the minimisation of squared "errors" forced the imported input to capture more of that progress than is its "fair" share. This would have the effect of biasing β upwards more than α and, consequently, of biasing also the ratio of the marginal product of imported capital to that of domestic capital upwards. This is, in fact, confirmed indirectly by Weitzman's tests, based on specification (9). He reports that adding the technical progress term does result, with the single exception of petroleum products, in minor or negative marginal productivity of imported Western capital.

The variation in the results reported by the studies is in part caused by the assumption concerning the degree of substitutability between domestic capital and imported machinery. This degree is expressed by the respective elasticity of substitution, which we denoted in Part A by σ_{KK}. The assumption concerning the value of this parameter is as follows:

Green-Levine $\quad \sigma_{KK} = 1$; relatively high substitutability;

Weitzman $\quad \sigma_{KK} = \infty$ in most cases tested; perfect substitutability. In some cases $\sigma_{KK} = 1$;

Gomulka $\quad \sigma_{KL} < \sigma_{KK} < \infty$.

We have already explained the meaning of these differences in the argument following equation (10). The constant returns to scale production function underlying Gomulka's estimates for Poland and the USSR, if expressed in aggregate rather than vintage-capital terms, would be of the type

$$Y = F(K_i, T_1 L_1) + F(K_d, T_2 L_2),$$

where L_1 represents employment on imported machinery and L_2 is employment on domestic capital, and where T_1 and T_2 represent respective levels of technology. Effectively, the economy is divided into separate parts, one consisting of imported capital and the other of domestic capital. The elasticity of substitution between capital and labour is assumed to be the same in both parts, but it may be constant or variable, high or low. The assumption (*i*) that the economy is separable into two parts and (*ii*) that the capital/output ratios, or the *average capital productivities,* are in both parts the same and constant together imply that σ_{KK} is both constant over time and greater than σ_{KL}[11]. These two assumptions are clearly both quite strong. However, the average productivity of labour is assumed higher on imported machines and the model is designed to trace the output implications of this particular difference. A mere shift in the composition of a given volume of machines, in favour of those imported from the West, would not immediately increase the output of the economy, but would release some labour which if, and only if, employed on additional machines (of whatever origin) would result in more output. Importation of technologically more advanced machinery by the USSR, apart from increasing the range of products and their quality, is thus perceived, in this model, to be basically a device to lift the labour constraint to growth. In the USSR and East Europe, labour is, more or less, fully employed. The level of employment being given, an increase in the share of Western capital would then result, in that model, in an increase in the level of output. This increase in output is called the direct effect of embodied technological diffusion.

The Green-Levine data for the Soviet Union suggest that the share of imported capital in the total capital stock of fixed assets has been relatively stable since 1960. If it were exactly constant over a long period, Gomulka's model would imply that the direct effect of machinery imports from the West has been to increase only the level of Soviet output, not its growth rate. But the direct effect is not the only one.

The growth rate could have increased due to the so-called indirect effect. This arises when the quality of the home-produced machines is improved just because some of them would be produced on the imported Western machinery (and technology). This is a potentially important effect, but its estimation is fraught with great difficulties. The empirical results reported by Gomulka are no more than of indicative nature, perhaps even not more than illustrations. As shown in Gomulka-Sylwestrowicz (1976) and Gomulka (1978), the particular assumptions adopted in that model imply that the indirect effect of machinery imports would increase the joint productivity residual, λ, by

$$\Delta\lambda = (1-\pi) \frac{I}{K} \frac{M_i}{M} \left(\frac{y_i}{y_h} - 1 \right) \qquad (14)$$

where π is the elasticity of output with respect to capital, I/K is the ratio of gross investment to the stock of capital, M_i/M is the ratio of imported machinery to all machinery investment and y_i/y_h is the ratio of the labour productivity on imported machinery to that on home-produced machinery. When the dynamic, or long term, contribution of this indirect effect is sought, we must take into account the impact of the indirect effect on the growth rate of capital. Following the argument of our introductory section where the measurement problem is discussed, it is clear that $\Delta\lambda/(1-\pi)$ is then the size of such long-term contribution to the growth rate of output, where $\Delta\lambda$ is given by (14). Coming back to the Soviet growth data, we note that the ratio I/K for the industry total has been falling significantly since 1960. At the same time the ratio y_i/y_h was probably stable and M_i/M was markedly increasing. The overall result is that the contribution of $\Delta\lambda$, as given by (14), would have been not only small, something like 0.2 per cent, but also stable over time. Taking $\pi = 1/3$, the contribution to Soviet industrial output growth of the imports of Western machinery would be then 0.3 per cent, or something like $1/20$ of the output rate of growth.

Is this quantitative result plausible in view of the (agnostic) results reported by Weitzman? It is not possible to give the answer to this question, since Weitzman's tests are not designed to explain the joint (capital-labour) productivity growth rate, λ, in terms of its constituent components to be attributable to different sources of technical change, of which technology transfer is one, but only to estimate λ as an aggregate. His main result, that marginal productivities of imported machines and of domestic capital may be not much different, puts into question the Green-Levine specification of the production function and their numerical estimates of the contribution of Western machinery imports to Soviet growth, but need not be at variance with the possibility, suggested by Eq. (14), that a part of the Soviet joint productivity growth rate may be attributable to the machinery imports from the West. The (approximate) equality of marginal products of both types of capital is not excluded in Gomulka's model. The actual computations of the model are made under the assumption of the same (and constant) average capital/output ratios whichever the capital. On purely efficiency grounds, which of course may not be respected in the USSR, one would, in fact, expect that production techniques and investment projects are selected by planners so that the two marginal products of capital are approximately the same. These two marginal products should be significantly different only if the marginal technology transfer costs are also large, and this may or may not be the case. It must, therefore, be stressed, in order to warn the reader against potential misinterpretation of the results of Weitzman's tests, that the *mere equality of marginal products* of both types of capital may be compatible (it actually is so in Gomulka's model) with the two types of capital being still qualitatively very much different, and with Soviet imports of Western capital contributing significantly to the growth of the Soviet economy. Weitzman's paper has not disproved this possibility because, we stress again, the rate λ is, in his model, assumed constant and independent of the size of machinery imports. His paper is important as a comment on the Green-Levine results, but of lesser importance as a study of the contribution of technology transfer to economic growth. The relationship between λ and the share of machinery imports in the total investment flow need not be particularly strong only when the absorbtive capacity of the importing country is binding, so that machinery is in part wasted, or when the actual relative technological gap between the importing country and the more advanced one is already close to its equilibrium level. In the latter case, the size of the research and technological transfer could be very large, the continuing importation of technology being indispensable just for keeping the *relative* gap at its constant level, and yet the rate of technological change would be, by past standards, both low and (almost) insensitive to increases in the size of the transfer. However, the ideas (*i*) that technological transfer makes a contribution to the *rate* of technological change, and (*ii*) that that contribution is a function of the technological gaps in various industries, have not been seriously modelled and tested by those who estimate jointly the diffusional effects and production functions using standard econometric methods.

Coming back to Gomulka's investment subvintages growth model, we note that his separability assumption implies no gains from the potential interaction between domestic and imported machines. This is perhaps realistic when turn-key factories are imported, but it is likely to be unrealistic in those, probably many, cases where the two kinds of capital co-operate in such a way that the productivity of one is affected by the presence and quantity of the other. Nevertheless, the distinctive feature of the model is its attempt to incorporate the ideas (*i*) and (*ii*) above.

Part D

IDENTIFICATION OF SPECIFIC PROBLEMS OF MEASUREMENT

1. LIMITATIONS OF AGGREGATE ANALYSIS

All the work discussed in Parts B and C is macro-economic in orientation, and most of it is based on time series analysis. Advantages and limitations of such macro-economic, or aggregative, analysis are well known and need not be repeated here. It may nevertheless be worthwhile to stress that aggregate level of technology is only a useful abstraction, perhaps even more abstract than aggregate capital or aggregate output. Technology is clearly a multidimensional "thing". But even if it were one-dimensional and measurable, the level of technology would probably differ from one piece of equipment to another. The product-mix of what is imported matters, and this varies over time. The composition of output and capital varies too, and these variations may not be uniform with time, producing oscillations in the residual term. The size of the data sample used by Green-Levine, Weitzman, Rosefielde and Toda is small (15 years), which renders the results of any attempted separation of the systematic influence from the random noise not very reliable. When no such separation is attempted, as in Gomulka's study, more specific assumptions were made instead, the realism of some of which is, to say the least, doubtful or uncertain.

Aggregative, or "macro", analysis necessarily involves a very large number of factors, each of which may have an effect on the outcome in ways which resist precise identification. We have already alluded to this problem in Part A. Apart from capital and manhours there are also other factors, such as improved training, better incentives to labour and management, a more positive attitude to work, a change in the relative weight of the more advanced industries (which would "improve" aggregate statistics even if no one industry had become more efficient), each of which may contribute significantly to the growth rate of aggregate output. Thus we cannot be quite confident of our measurement of the contribution of technological progress in general. Obviously the Western contribution accounts for only a part of the Soviet technological progress. However, it is particularly worth stressing that, as already indicated, only some of this contribution, possibly a small fraction of it, is embodied in Western machines that arrive in the USSR. There are indirect effects, through the learning process, involving designers, engineers, and workers, which affect the quality of Soviet-made machines, and which one has no means of measuring directly. We know that Western machinery imports represent only a small percentage of machinery installed in the USSR in the last twenty years. This means that the variation of the share of Soviet-made machinery in the total investment has been negligible. It is therefore not surprising that to identify confidently its separate contribution to technological change from year-to-year changes in output appears so difficult at present.

There are, of course, also problems encountered in industrial-sector and individual investment-projects studies, some of them similar to those listed above. But the microtype studies tend to identify with probably greater accuracy the direct costs and benefits of specific equipment to a given project or industry, and may also identify other influences on the productivity changes in that industry, such as new domestically-developed products and processes, the quality of the management techniques, the quality of the labour force, and so forth. Such studies, if carefully done, are however extremely time-consuming and yet liable to miss the indirect benefits, as has been repeatedly stressed in the Introduction.

2. THE CONTRIBUTION OF AN IMPORTED MACHINE IN A SHORTAGE-TYPE ECONOMY

At a microlevel, there might be instances when it is not possible to identify the contribution of imported machinery, even though all the facts are known. Instances of this kind may arise when several inputs, some of them of domestic origin, are used in fixed proportions to produce a single or joint product. Suppose production is constrained solely by a non-substitutable imported input, with some or all domestic inputs in excess supply. A relaxation of this constraint may easily result, then, in a startlingly large economic gain, which in part must reflect the contribution of the so far under-utilised domestic inputs. This (marginal) gain is what the economists call the shadow price of the imported input. In principle such prices could be calculated in all cases, also in those involving joint inputs. These short-term prices satisfy the short-term rule of free goods, by which any input in excess supply (domestic inputs in the example above) has a zero price. As a long-term alternative, the contribution of the imported inputs could be calculated as the value of output net of the resource cost of the domestic inputs, irrespective of their current value in terms of shadow prices.

The above argument is not only of academic interest. The Soviet is a disequilibrium-type economy in which excess supplies of some inputs and excess demands for other inputs are commonplace and persistent. In such an economy a recourse to swift and flexible imports is a powerful instrument of ensuring gainful use of these domestic inputs which happen to be in excess supply. Hence, there is probably a case here for imports to have a relatively greater effect than they would have had the USSR had a market-oriented economy. This point is related to resource trade rather than to technology transfer. That is to say, the import of any bottleneck item can have a substantial effect on output, by enabling complementary domestic inputs to be used effectively; the item could be pipe for the Siberian gas project, electronic components, spare parts for tractors, anything that is limiting the use of the inputs that are in surplus: it need not be a sophisticated machine at all. Moreover, the Soviet inventive activity is known to suffer from inflexibility and inefficiency as well. This activity tends to be concentrated in military-oriented areas, with many important branches, especially in the consumer sector, left deprived. One would expect large gains to be made from engaging in technology transfer, if the imported technology is directed above all to those deprived branches. However, systemic inflexibility reduces the efficiency of Soviet R & D activity so much that it often forces Soviet planners to make up for this by concentrating technology transfer in priority areas. Thus, the fact that the clothing industry is backward and imported Western machinery would probably "pay" large dividends in improved productivity and output there, it has yet to lead to a decision to import such machinery for that industry.

3. COMPLEMENTARY SOVIET INPUTS AND PRIORITIES

The measurement of the specific contribution of Western technology may thus be further complicated by the possibility that the observed greater speed of completion and efficiency in operation in certain projects based on Western technology is due in part to a high degree of priority of projects concerned, and only in part to a higher quality of Western equipment. Thus, for example, there is certainly a large contribution of Western technology in the chemical industry, and this industry's joint inputs productivity grew at a rate higher than average. But both this faster growth and larger import of foreign equipment may be seen as a consequence of the decision to upgrade the priority of the chemical industry, resulting in improved supplies and quality of the domestic inputs. In other words, while rapid productivity growth was facilitated by the imports, the increased degree of priority may also have been a contributing factor.

It should be noted that such changes in priority and the resulting transfers of high quality domestic inputs and construction capabilities from some investment projects to others need not have significant macro implications, but these clearly may bias the estimates of the contribution of machinery imports in specific projects or sectors, the direction of bias depending on the direction in priority change.

The same point has been made in the contribution by George D. Holliday[12]. He notes that "The resource-releasing aspect of transferring technology to the Soviet Union has received considerable attention in the West... (but) the resource-demanding aspect of technology transfer has received little attention. Yet there is evidence (he continues) that the resource-demanding effect is substantial. In some cases, the net result may be to draw resource away from traditionally high-priority sectors of the economy (such as the military sector)". Thus both the Volga (Tol-yatti) and Kama plants require not only direct high-quality Soviet inputs but also generate substantial indirect demands, e.g. for special kinds of metals, plastics, high-octane gasoline, new highways, service facilities and other infrastructures, in competition with other Soviet requirements. No attempt has been made in any sectoral studies known to us to take these factors into account in assessing the impact of imports on growth rates.

4. DISTINCTION BETWEEN EMBARGO EFFECTS IN SHORT RUN AND LONGER RUN

Denying the USSR some key item of equipment may cause disruption and a considerable loss in the short run. But if the item were considered essential, the USSR would no doubt attempt to develop it, and the ensuing research and development cost, increased by any cost associated with waiting for the item, would have to be compared with the potential transfer technology cost to determine the net loss in the longer run effect. This loss would no doubt vary from case to case, and may conceivably be very high in some instances. However, the thrust of the work on international technological diffusion and growth is that the policy of complete embargo for West-East technology trade would be capable of reducing the East's technological change and output growth significantly only if the Soviet and East European R & D sectors were to be sealed off from the Western R & D sector to a sufficiently high degree. The present almost free access to most of Western old and new science, as well as a somewhat less free, yet still relatively easy and inexpensive, access to Western old and new processes and product innovation, makes it possible for any highly industrialised country with large R & D resources, to develop or imitate at a relatively low cost most of the things which have been in use for some time, that is all except the "latest" technological vintage.

There is also the question of the Soviet planners being prepared, or unprepared, for the denial of Western technology. If Soviet planners know in advance that Western technology is not available, they would presumably alter their investment and R & D plans. Instead of counting on the technology transfer and the international division of labour, they would adopt an autarkic strategy. In this case the output loss due to denying the technology would clearly be lower than in the case when the transfer is stopped abruptly. Any estimate of the contribution (or the loss) to growth of the West-East technology transfer is therefore conditional upon the specification of Soviet expectations regarding the embargo policy.

5. THE QUALITY OF DATA ON IMPORTED WESTERN CAPITAL

Imported Western capital series for the USSR, constructed by Green, Levine and Jarsulic, and used in the Green-Levine and Weitzman studies, are built up out of import data. Underlying the series are two assumptions: (i) that the retirement rate is 5 per cent per annum and (ii) that the installation period is one year. However, micro-economic evidence suggests that the second of these assumptions might be quite unrealistic. Philip Hanson's recent survey of United Kingdom chemical-plant exports to the USSR finds that the lag between importing a machine and making the plant for which it was imported operational is in most cases between three and six years, with the average lag equal to about four years. The assumption of a much lower and constant time-lag of one year would have little impact on the estimated share of the stock of imported Western capital in the USSR. However, this assumption could have had, because of the volatile changes in the volume of imports from year to year, a serious impact on the estimated annual growth rates in that stock. Since the two studies mentioned above use data on these annual growth rates, the poor quality of the data might have had a major effect on the studies' estimates, especially because the sample is so small.

A one-year construction period is assumed also in Gomulka's study. However, in that study the important variable is not the growth rate of imported capital, but the share of that capital in total machinery investment. This share was relatively stable in the Soviet Union, and so almost immune to the length of the construction period, provided that it was more or less constant. In Poland, however, this share is changing sharply since 1971, and Gomulka's year-to-year estimates for that country would be affected seriously if the Polish (average) construction lag was nearly as long as the Soviet one.

Let us summarise the argument of this section of the paper. There are issues of aggregation, and of identifying the influence on the growth of output of technological progress in general, whatever its source, of separating it from other factors contributing to the productivity residual, and then of measuring the contribution of imported technology to the outcome. We note also that the measured contributions of technology imports may be biased when the degree of priority of the sectors to which imported equipment is destined changes, and with that priority changes also the quality of complementary domestic inputs and the speed of construction. Then there is the problem of distinguishing between long and short-run effects, in particular with reference to the disruption which can be caused by sudden and unexpected embargos or changes in plans. The poor quality of data on machinery imports to particular sectors, and also the possible changes in the time-lag between acquisition of equipment and its operational use, further complicate the task of researchers in this area.

There is also the fact that a part of the trade in investment goods does not involve any transfer of technology. Obviously the mere fact that a country imports some particular type of machine is not necessarily evidence of its technological backwardness in that particular area. In practice, however, it is not easy – in fact it seems impossible – to separate out the trade gains due to international specialisation at a given technology from the gains brought about by the import of more advanced technology. One may find consolation in observing that increased specialisation represents an organisational innovation and that it could, therefore, be treated as equivalent to a technology transfer; it is only "equivalent to" because no outside technology is involved, but it is a transfer of sort because the innovation cannot be adopted without the consent of the outside partner.

Part E

SYSTEMIC FACTORS IN LOW ECONOMIC EFFICIENCY UNDER SOVIET-TYPE CENTRAL PLANNING AND MANAGEMENT

Soviet books, periodicals and speeches by leaders, refer repeatedly to the unsatisfactory state of Soviet innovative and investment activities: high costs, delays, wrong technological choices, the insufficient quantity and poor quality of machinery available, slow introduction of new products and processes; the insufficient availability and training of labour, the poor organisation and planning of machinery production, are all subject to severe criticisms. Efforts are being made to overcome the defects which will be further analysed below. The keys to Soviet efficiency, including the efficient use of imported technology, are clearly internal; it cannot be otherwise. They are to do with the nature and deficiencies of centralised planning. That these deficiencies exist and are all important would be accepted by any serious Soviet economist, and probably by the Soviet leaders too. Any argument would be about their relative importance, and perhaps also about causes and remedies.

i) *Excessive duration* of decision-making, construction, installation, "running-in", of new productive capacity. There is an elaborate and clumsy hierarchy of "project-making organisations", whose plans are often expressed in terms of the value of the projects they draft, so they can be penalised for devising an economical, low-cost project! There is a big gap between technological research and the adoption of its results by the investment decision-makers and by the producers of machinery and equipment. Many Soviet sources testify that by the time the factory becomes operational – which may be 10-15 years after the original decisions were taken – the "new" machines it contains are already obsolete (and this includes imported machines). A major cause of delay appears to be the so-called "scattering of investment resources among too many projects", a disease publicly criticised for many decades but apparently incurable.

ii) *Insufficient interest in innovation* on the part of management. This has multiple causes. Thus risk is seldom rewarded, yet any innovation carries with it a risk of failure. Management's attention is usually centred on fulfilling current plans, but innovation could have a short-term disruptive effect. New designs, new models, must be "fitted" into aggregate plan targets, measured in money terms, tons or other such units. A manager desirous of actually taking a risk and showing initiative is hampered by the clumsy system of allocation of materials and of investment funds by superior administrative bodies, and there are few innovations which do not involve some change in material inputs and/or investment funds.

iii) *Slow retirement of obsolete equipment.* This has several causes. One is the tendency to build new factories rather than modernise existing ones. [This tendency, incidentally, reduces somewhat the negative effects of the cause (*ii*) above.] Another is the pressure to maximise output from any existing capacity with little regard for cost, this in turn being due to chronic shortages. This tendency impedes maintenance. The recent fall in output of iron and steel has been attributed to prolonged neglect of the need to replace worn-out equipment, due to over-full use of capacity. Instead of being replaced, old machines are repeatedly repaired, patched up, often at very high cost. This practice is also due to shortage of new machines. A Soviet source suggests that 40 per cent of all machine-tools, and millions of skilled workers are used to keep obsolete and unproductive capital equipment in some sort of working order[13].

iv) *Neglect of malaya mekhanizatsiya* ("small" mechanisation, i.e. of auxiliary tasks, such as loading, materials handling, packaging) which creates labour bottlenecks and reduces the effectiveness of "basic" machinery. Planning organs and ministries tend to concentrate on large-scale projects, with neglect of complementary tasks. These are often given, as sidelines, to large enterprises whose primary attention is devoted to other items, and there are few small enterprises, with a continued tendency towards mergers. Small and uncomplicated machinery does not rate highly in plan fulfilment targets.

v) *Shortage of specialised producers and self-supply.* The best and most forthright criticism of this appeared in two articles by S. Kheinman (in *Ekonomika i Organizatsiya Proizvodstva*, Nos. 5 and 6, 1980). The author divides machinery and engineering production into three categories: that produced by specialised "machinery" ministries, that produced by specialised plants serving the needs of non-machinery ministries, and that produced in workshops in non-machinery enterprises, usually for their own use. Good quality, efficient production and economies of scale are much more likely in the first of these three categories, least likely in the third. There is a very serious lack of specialised production of components, tools, instruments, spare parts. Many of these have to be made for own use ("self-supply") by enterprises of all kinds which cannot rely on supply from outside sources. Production has to be improvised on unsuitable premises and with the aid of inadequate equipment, often in uneconomically small batches. This, and the practice of repeated repairs to obsolete equipment, plus a virtual absence of after-sales service by manufacturers of machinery, leads to massive waste of manpower and the misuse of capital which, Kheinman states, is why the USSR has more machine-tools than the United States, West Germany and Japan together. Many machines are made as "sidelines" by all sorts of enterprises: thus materials handling equipment is made by enterprises within no less than 35 different economic ministries. This may reduce economies of scale, inhibit technical progress and prevent standardization of parts. Soviet sources often point to the contrast in productivity and quality between specialised production and often primitive "self-supply" workshops.

vi) *The effect of "success indicators" on machinery production.* As is well known, Soviet managers have to fulfil plans, which are expressed in some aggregate units of measurement: rubles, tons, number of units, and so on. It has proved particularly difficult to incorporate in such aggregate plan indicators the detailed requirements of the users. "For decades the gross value of output and tons were dominant indicators. To this we owe the fact that our machines are often heavier than similar foreign ones... Managers, facing difficulties, choose the product variant which causes least trouble for themselves, which usually means attention to quantity at the cost of quality" (*Pravda*, 7th May, 1982). "The producer has virtually no responsibility for the effectiveness in use of his product... Deficiencies have to be put right by the user" (*ibid.*). This is an aspect of a long-standing disease of the system: consumer (user) demand has far too little influence on production. Instances have been reported in which Soviet managers seek to obtain an imported machine merely because they then stand a better chance of obtaining the one they actually require. A number of articles have been published in which Soviet managers of machinery plant complain that the success indicators (expressed in rubles and tons), and also plans imposed for economy of materials, prevent them from producing what their customers actually need. The chronic shortage of spare parts is also partly due to plan fulfilment indicators: they count for less than complete machines.

vii) *Rising costs.* Numerous critical articles point to the fact that machines are becoming disproportionately dearer, and there is also an "inflation" in construction costs. Pressure to reduce costs is ineffective, and the "reduction" in machine prices which is claimed in the official price index is seen by Soviet critics as quite misleading (see for example V. Krasovsky, in *Voprosy ekonomiki,* No. 1, 1980, and K. Val'tukh, *EKO,* No. 3, 1982). The state plans provide for increased allocation to investments in "real" terms, but, according to these Soviet economists, the volume of investment may actually be declining, contributing thereby to shortages and to the need to retain obsolete productive capacity. High cost is unintentionally stimulated by the success indicators: thus construction has hitherto been planned in terms of rubles *spent,* and dearer machines contribute more to aggregate ruble targets of output value.

viii) *The absence of competition* and *price inflexibility* cause chronic shortages, reduce the influence of the customer, limit choice, induce a take-it-or-leave-it attitude. The customer is tied to a particular supplier by the allocation system. If there is no competition between suppliers, shortages create an atmosphere in which the customers have to compete for limited supplies, with all the consequences that follow. Incentive is lacking to please the user, to retain his goodwill, to provide after-sales service. A *Pravda* correspondent noted the value of competition for the defence industries, "where there takes place a continuous and inescapable comparison with foreign technology".

ix) *The deficiencies of the material supply system* have an adverse effect on machinery manufacture (e.g. difficulties in getting steel of the correct specification), and also on the operations of installed machines.

This basically unfavourable environment limits the effectiveness of imported technology too. According to Philip Hanson, "the domestic diffusion of technologies imported from the West is a particularly slow and problematic process in these Eastern economies" (*Journal of Comparative Economics,* No. 6, 1982, p. 158). The desire of the leadership to stimulate and accelerate the diffusion of new technology, domestic and imported, is clear and repeatedly stressed in Party documents. However, these intentions are frustrated by the clumsy malfunctioning of an overcentralised and bureaucratised system of planning and management. But the needed reforms run counter to habit, vested interest, Party traditions and the formal ideology, which is hostile to handing over resource allocation to the market mechanism and resists (even) Hungarian-type reforms. The remedies so far proposed are based on strengthened centralisation, and so are most unlikely to succeed. The contribution of Western technology to the correction of these serious defects of the system can only be quite minor. The needed changes are essentially internal.

Part F

THE FAILURE
OF THE IMPORT-LED GROWTH STRATEGY

Among the six Eastern European countries the strategy of import-led growth was initiated first by Roumania in the mid-1960s, to be then followed by all the remaining five countries in the 1970s. Particularly strong enthusiasts for the strategy have been Hungary and Poland. The economic rationale of the strategy is discussed by Gomulka and Sylwestrowicz (1976) and in Gomulka (1978). Philip Hanson sums it up in the following way:

> "Imports of technology, in both embodied and disembodied way, would play a major part in improving the productivity and technological level of the Polish economy; in contrast to earlier Polish policies, the convertible-currency import bill would be allowed to rise substantially faster than convertible-currency revenues, and a large current-account deficit would for a time be financed by borrowing from the West, facilitating an acceleration of investment growth in the medium term without putting great pressure on consumption; but the modernisation of Polish production would meanwhile be transforming Polish capabilities for exports to convertible-currency trade partners, or for import savings, or both; and some combination of hard-currency export growth and production that would save hard-currency imports would, in due course, contain the growth of external debt and allow the process to continue without financial constraints forcing a halt." (1982, p. 121)

If the strategy was to succeed it was essential to build up the export capability sufficiently large to sustain for a long time the increased share of Western machinery imports. This has not been done and consequently the strategy failed.

The studies by Zdenek Drabek (1981) and Philip Hanson (1982) have attempted to identify the major reasons for this failure. They both find little correlation between the size (or the growth) of technology imports by industrial branches and the corresponding indicators (again, either size or growth rate) of the branches' dollar exports. Moreover, about 30 per cent of Western technology was absorbed by sectors producing non-traded goods (Drabek, p. 47). A temporary relaxation of the balance-of-payments constraint produced, in the 1970s, eagerness to import also technologically simple intermediate inputs and consumer goods. Imports of the latter category have been particularly large in Poland, bringing the payments crisis forward, before major projects of the modernisation programmes could have been completed. There are thus indications of serious mismanagement of the strategy at the top, especially in Poland. Perhaps similar type mismanagement has occurred in some market-oriented economies as well, such as Argentina, Brazil or Mexico. However, the distinguishing feature of the centrally planned economies remains their relatively low capability to export manufacturing goods, and this feature can in turn be traced to the peculiar institutional environment in which their individual enterprises operate; attention to some important aspects of that environment has been given in the Part above.

Part G

CONCLUSION

We take the view that the Green-Levine estimates of the contribution of Western capital goods to Soviet growth cannot be regarded as reliable. Using the same data but different methods, Weitzman and Toda obtained results which suggest that the influence of Western machinery/imports on Soviet growth is either small or uncertain. Also the theoretical analysis and empirical calculations due to Gomulka point to the relatively small importance of machinery imports as a factor in Soviet economic growth, especially in the postwar period. However, this work suggests that other forms of science and technology transfer have probably played a very important role. The contribution of imports of Western machinery to growth would appear to have been significantly greater for East European countries, whose relatively small R & D sectors make for smaller inventive/absorptive capacity and who have always imported relatively much more from the West, especially in the 1970s.

This may seem to be a very general and imprecise statement, but in our view the data we have at the moment and the results of the other calculations and arguments do not permit a more precise evaluation. Perhaps a large number of detailed studies of particular industries and investment projects may jointly offer a better chance of identifying more accurately the aggregate costs and benefits of technology transfer.

We would also like to draw attention to the hypothesis that the extent of gains from technology transfer in terms of the contribution to the *growth* rate of output is, up to a certain point, related positively to the size of the *relative technological gap*. Such a relative gap is indicative of the opportunities for technological innovation which have not yet been taken up. Both theory and evidence suggest that the exploitation of these opportunities underlies the more rapid growth of all those countries which have either foreign exchange to import machinery and know-how on a large scale or sophisticated R & D sectors to be capable of assimilating large chunks of foreign technology. The USSR is probably closer to the latter category of countries. But this hypothesis implies also that success in reducing the relative technology gap would, at some point, initiate a decline in the innovational rate, bringing it down to that enjoyed by the areas such as the United States representing the world's technological frontier.

The Soviet innovational rate has been, it seems, declining over the last ten years or so, and now does not seem higher than the (rather low) United States rate. But the technological and productivity gaps remain still quite considerable. Are these, then, to be the Soviet Union's equilibrium (or minimum) gaps, given the present economic system? If so, any further decline of the gaps could be brought about not so much by increased technology imports, but by systemic changes intended to sharply increase the local competitive pressures and incentives to assimilate the better products and processes. In our view, improvements in this system area require wide-ranging internal reforms in the direction of more competitive, market-oriented, economic relations. But such reforms are, in the USSR, unlikely to be adopted, not in the near future anyway. For the time being we would, therefore, anticipate continued Soviet and East European reliance on the transfer of Western science and technology, especially in its disembodied form, without however the East making much further progress in closing the present technological and productivity gaps at the aggregate level; we expect improvements in some sectors to be cancelled out by deterioration in other sectors. The experience of the Soviet and East European countries in the 1970s has exposed vividly the weaknesses and the limits of the strategy of import-led growth. The prevailing Western view that such a strategy cannot be really effective, at this rather advanced stage of Soviet and East European development, in closing further the technological gaps between them and the West without massive opening up of their economies to market competition and financial discipline has apparently been vindicated by that experience. It is interesting that probably most East European economists, especially Polish and Hungarian ones, would now, after what happened in the 1970s, go along with that Western view as well. It may be noted that the imports of machinery and expertise from the West have almost collapsed in the 1980s, primarily under the weight of the rapidly rising debt. It seems that disillusion – a feeling that increased Western technology imports alone cannot pull their economies up – must also have been a contributing factor.

The responses of the communist elites, in terms of economic reforms and foreign trade policy, may differ significantly between those of the USSR and those of the East European countries. The USSR does not need to import Western machinery on a large scale to take good advantage, although admittedly with some delay, of Western innovative activity. The pressure to decentralise there is therefore weaker than, say, in Poland or Hungary, where the domestic R & D activity is too insignificant to be an effective instrument of international technology transfer and where, consequently, larger export capability must be developed instead to sustain larger embodied technology imports. By the same argument, the reform designed to enhance economic efficiency and trade competitiveness would become less pressing for Eastern Europe as well, should the West impose an embargo on technology exports to this region. Should such an embargo prove to hold successfully, Eastern Europe might consider moving towards technological integration with the USSR. It may be assumed that any attempt of this kind would be unpopular among the population of the region and perhaps not very attractive for the USSR either. In any case, Eastern Europe is the weak spot of the Soviet Bloc and is likely to suffer most from any determined policy of the West to reduce substantially the West-East technology flow.

NOTES

1. The authors wish to thank Alberto Chilosi, Donald Green, Philip Hanson, Martin Spechler, Yasushi Toda and, in particular, Mrs. Helgard Wienert from the OECD, for their helpful comments and advice.
2. Imogene Edwards, Margaret Hughes, and James Noren, "US and USSR: Comparison of GNP", in *Soviet Economy in a Time of Change – A Compendium of Papers*, Joint Economic Committee, Congress of the United States, 1st Session, 10th October, 1979, p. 377.
3. *Ibid.*
4. Louvan E. Nolting and Murray Feshbach, "R&D Employment in the USSR–Definitions, Statistics and Comparison", in *Ibid.*, p. 747.
5. That is, the imports would increase the industrial growth rate from, say, 4 per cent to 5.1 per cent.
6. Note the α in this section is not the same as α in Section A.2.
7. For example, if the (equilibrium) growth rate of the machine building and metal working output were 5 per cent, the drop would be 0.26 × 5 per cent = 1.3 per cent.
8. That is, the imports would increase the industrial growth rate from, say, 4 per cent, to 4.3-4.6 per cent for the USSR and 5-7 per cent for East Europe.
9. We recall that this statement is supported first by the theoretical argument and empirical evidence of the type referred to above indicating that diffusion of outside scientific and technological innovation has been a key factor in Soviet growth, and, second, by the empirical evidence due to Gomulka, Rosefielde and Weitzman, indicating that one particular form of that diffusion, namely that through machinery imports, is probably of minor significance. The burden of significance is therefore shifted to other forms of diffusion. It must be stressed though, that the arguments underlying the statement in question are not free of question marks.
10. According to the econometric estimates reported in Gomulka (1977), the programme to reduce the number of working hours per worker, which was implemented in the years 1956-61, produced a disturbance: it led to an increase in the productivity residual in those years and a decline in the following several years.
11. It may be shown that if F is assumed to be a constant-returns to scale and a constant-elasticity-of-substitution production function, then $\sigma_{KK} = \sigma_{KL}(1 - \pi)$, where π is the imputed capital share. Because technical progress is assumed to be neutral in the Harrod-Kalecki sense, it follows that both the marginal product of capital and the imputed capital share depend only on the capital/output ratio, and so they would take on the same values in both parts of the economy, if the capital/output ratios were the same. The important assumption of the same capital/output ratios is not based on any direct evidence, but on the fact that the average aggregate capital/output ratios in manufacturing sectors appear to be quite similar among major industrial countries and stable over time. Despite this fact, the assumption could well be widely unrealistic in specific instances.
12. George D. Holliday, "The Role of Western Technology in the Soviet Economy", in *Issues in East-West Commercial Relations – A Compendium of Papers*, Joint Economic Committee, Congress of the United States, 95th Congress, 2nd Session, Washington, D.C., 12th January, 1979, p. 47.
13. See Appendix 1 for sources.

REFERENCES

Amman, Ronald; Cooper, Julian and Davies, Robert (eds.), *The Technological Level of Soviet Industry*, Yale University Press, 1977.

Berliner, Joseph S., *The Innovation Decision in the Soviet Industry*, MIT Press, 1976.

Desai, Padma, "The Production Function and Technical Change in Post-war Soviet Industry: A Re-examination", *American Economic Review*, Vol. 66, June 1976.

Gomulka, Stanislaw, *Inventive Activity, Diffusion and the Stages of Economic Growth*, Aarhus, Denmark, 1971.

Gomulka, Stanislaw, "Import of Technology and Growth: Poland 1971-1980", *Cambridge Journal of Economics*, March 1978.

Gomulka, Stanislaw, "Industrialisation and the Rate of Growth: Eastern Europe 1955-75", *Journal of Post-Keynesian Economics*, Spring 1983, Vol. V, No. 3.

Gomulka, Stanislaw, "Slowdown in Soviet Industrial Growth 1947-1975 Reconsidered", *European Economic Review*, 10, 1977, pp. 37-49.

Gomulka, Stanislaw and Sylwestrowicz, Jerzy, "Import-led Growth: Theory and Estimation", in Franz-Lothar Altman, Oldrich Kyn, Hans-Jürgen Wagener (eds.), *On the Measurement of Factor Productivity: Theoretical Problems and Empirical Results*, Vandenboeck and Ruprecht, Göttingen, 1976.

Green, Donald W. and Jarsulic, Marc, "Imported Machinery and Soviet Industrial Production, 1960-1974: An Econometric Analysis", Soviet Econometric Model Working Paper No. 39, September 1975.

Green, Donald W. and Levine, Herbert S., "Implications of Technology Transfer for the USSR", NATO, Economic Directorate, *East-West Technological Co-operation*, Brussels, 1976.

Green, Donald W. and Levine, Herbert S., "Soviet Machinery Imports", *Survey*, Vol. 23, No. 2, Spring 1978, pp. 112-126.

Green, Donald W. and Levine, Herbert S., "Macro-economic Evidence of the Value of Machinery Imports to the Soviet Union", 1977, mimeo.

Green, Donald W. and Higgens, Christopher I., *SOVMOD I: A Macroeconometric Model of the Soviet Union*, Academic Press, 1977.

Hanson, Philip, "The Import of Western Technology", in Archie Brown and Michael Kaser (eds.), *The Soviet Union Since the Fall of Khruschev*, Macmillan Press Ltd., London, 1975.

Hanson, Philip, "International Technology Transfer from the West to the USSR", in *Soviet Economy in the New Perspective*, Joint Economic Committee of the United States Congress, Washington, D.C., October 1976.

Hanson, Philip, "The Impact of Western Technology: A Case Study of the Soviet Mineral Fertilizer Industry" (mimeographed), University of Birmingham, 1977, forthcoming in Paul Marer and John Michael Montias (ed.), *East European Integration and East-West Trade*, Indiana University Press, 1978.

Hanson, Philip, *Trade and Technology in Soviet-Western Relations*, Macmillan Press Ltd., London, 1981.

Hanson, Philip, "The End of Import-Led Growth? Some Observations on Soviet, Polish and Hungarian Experience in the 1970s", *Journal of Comparative Economics*, Vol. 6 (2), June 1982.

Laski, Kazimierz, "Capital and Equipment Imports and Growth in Socialist Countries" (mimeographed), Linz University, 1979.

Moodie, Carlise and Rushing, Francis W., "Technological Change in the Soviet Chemical Industry", Report for the Strategic Studies Centre, February 1975, mimeo.

Rosefielde, Steven, "East-West Trade and Postwar Soviet Economic Growth: A Sectoral Production Function Approach", Report prepared for the Strategic Studies Centre, June 1978, mimeo.

Slama, Jiri, "Technologietransfer zwischen Ost und West in den 70er und 80er Jahren", *Ost Europa-Wirtschaft,* June 1982.

Spechler, Martin C., "Soviet Policy towards Technological Change since 1975", Tel-Aviv University, 1982, mimeo.

Sutton, A.C., *Western Technology and Soviet Economic Development,* Stanford, 1973.

Toda, Yasushi, "Technology Transfer to the USSR: The Marginal Productivity Differential and the Elasticity of Inter-Capital Substitution in Soviet Industry", *Journal of Comparative Economics,* June 1979.

Trzeciakowski, W. and Tabaczynski E., "The Impact of Technology Transfer on Economic Growth", in *Industrial Policies and Technology Transfer,* C.T. Saunders (ed.), Vienna, 1977.

Usher, Dan, *The Measurement of Economic Growth,* Blackwells, 1980.

Weitzman, Martin L., "Soviet Post-war Economic Growth and Capital-Labour Substitution", *American Economic Review,* September 1970.

Weitzman, Martin L., "Technology Transfer to the USSR: An Econometric Analysis", MIT Press, 1978; *Journal of Comparative Economics,* June 1979.

Zaleski, Eugene and Wienert, Helgard, *Technology Transfer Between East and West,* OECD, Paris, 1980.

OTHER REFERENCES

"Soviet Chemical Equipment Purchases from the West : Impact on Production and Foreign Trade", CIA, National Foreign Assessment Centre, 1978.

Appendix 1

A SAMPLE OF QUOTATIONS FROM SOVIET OFFICIAL SOURCES ON THE DEFICIENCIES OF SOVIET INNOVATIONAL PROCESSES

From V. Trapeznikov, *Pravda*, 7th May, 1982

"A reduction of return on capital took place. During 1958-1980 it fell from 0.48 to 0.31. The return on new investments in 1958 was 0.58, but in 1980 it fell to 0.16. Thus, if in 1958 an increase of productive capital by one ruble brought in 52 kopecs increase of national income, in 1980 the figure decreased to 16 kopecs. These figures should alarm us. They are not accidental, bearing in mind that insufficient tempo of replacement of obsolete machinery and equipment by more progressive ones in some branches of industry doomed these branches to physical and moral ageing.

"In fact the planned coefficient of retirement for basic production assets in industry was 5.6 per cent a year, which corresponds to their complete renewal in 18 years. But this coefficient lately tended to decrease and the period of renewal tended to increase.

"There are several reasons explaining this phenomenon, but they are all rooted in underestimation of scientific-technical progress and insufficient attention to developing new machinery. As a rule scientific achievements are realised through the use of machinery and instruments. However, planners overlooked longer-term considerations, and as a result a very low share of the total investments was allocated to these branches of industry. On the other hand, amortization and other assignations are wrongly allocated at the industry enterprises. Very little finance is assigned to replacement of equipment. As a result the majority of workshops – during some periods up to 40 per cent – are engaged in repairs of equipment instead of directing resources towards the creation of new and modern machinery.

....

"In my opinion when considering the importance of scientific-technical progress and its inadequate rate one should judge the measures on improvement of economic system primarily by the following criterion: whether they will accelerate or slow down technical progress; whether they will change the existing situation when obsolete production is profitable for manufacturing enterprises, but production is not.

....

"A wrongly chosen criterion, acting in thousands of localities of the country over a long period, brings about undesirable results in the end. For decades we adhered to such notorious indications as "gross-output" and "tons". We can blame on this the fact that our machinery and equipment are much heavier than foreign machines of the same type. An example of utilisation of contradictory criteria is the control of quality and quantity of products. In either case a need arises for additional labour and materials. How does a manager act when dealing with difficulties to fulfil plans? He usually chooses the alternative which causes him minimum troubles and preferably the output of larger quantity of goods at the expense of their quality. Nothing compels him to decide otherwise, although society is often more interested in quality.

....

"We should attach great importance to the export of our goods to markets in the countries with progressive technology. That is a sure way of objective verification of quality of our production. It is high time to evaluate the activity of Research Institutes and design offices in our country by comparison with that of the leading Research Institutes and firms abroad.

"Another requirement of the theory of management is feedback, that is the influence of successive links of the system of management on the preceding ones. In the national economy it is the influence of the consumer on the producer – which is so far insignificant – and the effect of management upon another manager, etc. In our national economy feedback is extremely weak and acts with great delay, and because of shortage of some products it practically disappears, and then the following rule comes into effect: take it or leave it; if you do not, others will.

"Machinery manufacturers are in fact not responsible for the effectiveness in use of their machines. The transaction once completed, a manufacturer disappears and a user has to correct defects himself, very often using primitive methods. The manufacturing enterprises should set up service units which deal with defects at their expense, and this is bound to stimulate the increase of quality of production. In many countries any big firm has service departments; with us they are in embryo so far. To start such service is to improve the feedback in the chain: consumer-manufacturer.

....

"A clear requirement of the theory of management is to take the human factor into account. People are present in every system of management. A human being is an "active system" with certain wishes and aims. There are various incentives which can direct aspirations and ambitions of different people in useful ways for national economy. One of them, very rarely met, is competition, which provides the basis for correct choice. Constant comparability of quality of goods with that of other domestic and foreign firms is an effective moving force of progress. This is especially apparent in the defence industry where there is continuous and inevitable comparison with foreign technology, which stimulates the maintenance of high scientific-technical levels.

"Sometimes the opposite way is chosen. A particular organisation becomes a monopolist in its field. The grounds for this is the following principle: let us concentrate everything in one place. However, monopoly usually hinders technical progress and in the end it suppresses any scientific-technical initiative. In this case there is no choice and a consumer has nothing to do but accept what is imposed on him; an organisation-monopolist gradually refuses any active search for new decisions and exists by the rule: if my production does not suit you, do it yourself. Thus lethargic and lazy attitudes are being cultivated and naturally technical progress slows down."

From G. Kurbatova, *EKO*, No. 3, 1982

"There are two ways of renewal of fixed capital assets. They are the following: addition of new machines to the existing park, and replacement of obsolete and worn out machines by new ones. Until now the former has prevailed in the national economy, i.e. the capital reproduction went on mainly at the expense of the increase of the park, but not enough attention was paid to replacement and renewal. In industry the share of equipment intended for replacement was 19.5 per cent in 1966 and 25 per cent in 1975. As a result the available equipment, being already economically and physically worn out, was not replaced by new ones but was often restored thanks to three or four and sometimes to more major repairs. According to estimates by some specialists, only the first such repair of equipment is economical; with successive repairs the efficiency of using obsolete equipment constantly decreases. This way of reproducing fixed capital assets has become one of the reasons for their ageing and the decline of the growth tempos of the industrial production and, hence, for the slowdown of the technical progress.

"The slow renewal of the equipment park has led to the predominance of obsolete equipment in some sectors. Thus, at present in ferrous metallurgy, more than half of rolling mills are obsolete, non-mechanised units; labour productivity is much lower there than with the new ones. The modern rolling mills provide only 45 per cent of rolled metal. According to specialists, the renewal of only one-tenth of obsolete rolling mills by modern ones would give about 14 million additional tons of rolled metal. Similar examples exist in other branches of industry. The need to raise our engineering to much higher levels has become essential.

....

"We have had disproportion in the volume and the rate of development between, on the one hand, engineering and metal-working and, on the other hand, those branches of industry which provide the inputs for the engineering industry.

"In the first place this concerns ferrous metallurgy, which is the main supplier of materials for engineering and metal-working. The machine and metal-working industries consume 2/5 of rolled ferrous metals. Therefore the low and declining growth rates of ferrous metallurgy affect the development of engineering. In addition the assortment and quality of our ferrous metals do not fully answer the requirements of engineering, which also hinders its progress and limits the efficiency of its production.

"Such a situation is explained by the lack of balance between ferrous metallurgy and engineering, particularly metallurgical machinery, whose level does not correspond to the tasks of the technical re-equipment of metallurgy production. The trouble is that specialised metallurgical engineering virtually does not exist in our economy. It is mainly represented by plants which manufacture not only metallurgic equipment but a wide range of equipment for dozens of different branches of industry.

"The proportion of investments in the metallurgic machine industry in relation to the capital investments in ferrous metallurgy is not large, and in the past five-year plans it tended to decrease. And if we take into account the fact that this branch provides the equipment for non-ferrous metallurgy and export deliveries, the share of capital investments in the production of machines and equipment for ferrous metallurgy will turn out to be even smaller.

"The existing capacities of metallurgical machinery industries cannot provide for the development of new progressive technological processes in metallurgy.

....

"As is known, the basis of engineering production is metal-working equipment which has been developed only recently in our country. However, the proportion of metal cutting machines and forging and pressing equipment, which have been in use for less than 10 years is constantly decreasing. This is because most of the annual deliveries of metal-working equipment are added to the existing park, whereas the annual volume of replacement stays at the level of the fifties.

"The coefficient of renewal, reflecting the interrelation of accumulation and renewal of machinery, is lower in our engineering than that in industry as a whole and is tending to fall. In 1960 it was 0.25, but in 1975 it reduced to 0.20 – on the average in industry as a whole this coefficient was equal to 0.25 in 1975. The share of annually renewed equipment in the total park of the machine and metal-working industries did not exceed 3.0 per cent during 1966/75 – with the exception of the year of 1972 when the rate of renewal of the equipment in this branch was equal to 3.23 – which is obviously not enough for the intensification of production. According to specialists, the proportion of the annual renewal of the metal-working equipment should not be smaller than 6-8 per cent.

"True, in some branches of engineering, the rate of renewal of the equipment is substantially higher than the average for the entire industry. Thus, in instrument-making it was 5.1 per cent in 1975; in machinery it was 5.96 per cent; and in the lathe and tool industry it was 7.52 per cent.

"However, as the renewal of the equipment is not carried out at the plant in a comprehensive manner, i.e. only separate machine-tools or items of equipment are replaced but not entire technological lines, it does not have the proper effect. Besides, in some branches of engineering and particularly in metallurgical and in materials-handling equipment, less than 1 per cent is renewed yearly.

"Therefore in industry, along with modern aggregates, which represent the latest achievements in technique and science, obsolete machines and equipment are used. On the one hand, obsolete machines divert a large proportion of the work-force, which leads to the decrease of application of new technology and hence, to substantial loss in the productivity of labour and machines, excessive use of metal and deterioration of production quality. Also the obsolete equipment and machines cause over-expenditures on major repairs."

....

From S. Kheinman, *EKO, No. 5,* 1980

"The widespread lack of balance in the plans, which amount to "planned shortages", inevitably cause the users to grab all they can, and eliminates all stimuli for the producers to take user needs into account" (p. 40).

"...A very large and growing part of [engineering] resources are used not in conditions of concentrated and specialised production of similar products, not in specialised sectors with appropriately developed 'production culture' but in 'their own' enterprises under a large number of ministries, in 'their own' auxiliary workshops in many sectors of the economy, i.e. in surroundings quite alien to the production of the items concerned.

"There are at present three kinds of engineering [machine-building]: engineering works within engineering ministries, engineering works within non-engineering ministries; and, thirdly, mechanical and repair workshops in non-engineering enterprises. Only just over half (55 per cent) of all machine-tools in enterprises of the first two categories...

"The above-named "second" engineering is, by the logic of its position, outside of any unified technological policy, outside technical progress and its organisation. The ministries concerned (e.g. Coal, Metallurgy, etc.) do not provide technological guidelines, they are customers...

"As a result of the existing organisation of the production and repair of machinery, there are engaged in this about 20 million persons and a larger number of machine-tools and forging and pressing machines than in the United States, Japan and West Germany combined... The 'law' of self-supply, the striving to provide 'one's own', is stronger than any criteria of effectiveness. Thus we know that our large car factories make their own large and rather complex machine-tools (including castings for them)... This kind of self-supply is considered to be consistent with the scientific-technical revolution. How can one improve the quality of machinery on such a basis!

"The practice continues of scattering the production of many basic and auxiliary machines and components among many ministries and enterprises. This shows lack of product specialization. Twenty-two factories in the Ministry of Heavy Machine-building produce 17 per cent of materials handling equipment, the rest being scattered among 400 factories within 35 different ministries. 155 factories in the Ministry of Road-building machinery make 82 per cent of road-making and communal equipment, the rest is made in 400 enterprises of other sectors... Almost every engineering works makes toothed wheels for their own use, 65 per cent of all reinforcing metal parts are made for and by themselves. All this in small quantities and at high cost.

"Centralised (specialised) production of components does not exceed 3 per cent, of mouldings about 4.5 per cent. There are at present no specialised factories for forging and castings in the USSR; these are the task of some 6 000 enterprises and workshops...

"We have in practice no engineering enterprises or corporations which produce non-standardized special-purpose equipment and technology where user-enterprises, or research institutes, can place orders. These needs are met ad hoc by the chief mechanical engineer's or other departments of the users, expending vast quantities of equipment, materials and labour.

"The first category of engineering is run by over 20 all-union ministries which have no territorial organs. In addition there are several dozen non-engineering ministries which control enterprises in the so-called second category of engineering. The various workshops which make up the third kind of engineering are not co-ordinated and are scattered among thousands of enterprises in almost every ministry. With the exception of Gosplan of the USSR, there are no State organs which co-ordinate all these in its regional aspect, and there is not even the prospect of the creation of such organs, although it has been repeatedly proposed. Also absent are any general institutes for the technology organisation and economy of engineering (machine-building) as a whole. Furthermore, the very principle on which engineering ministries are based requires critical attention."

From S. Kheinman, *EKO, No. 6,* 1982

"A major defect in the structure of our engineering industry is the relatively small share of specialised production of instruments and technological equipment. Despite numerous decisions, most kinds of machines are not supplied with sets of replacement parts, with hardly any specialised production of units required for the modernisation of machinery. As a result, the various sectors which use the equipment have to make their own spare parts and castings for repairs and replacements, using obsolete and ineffective methods.

"Our engineering industry and servicing units absorb an immense volume of labour and material resources. In 1978 about 40 per cent of all workers and over half of all engineer-technical staffs were employed in engineering and metal-working... Several million persons are engaged in repair-and-mechanical work in non-engineering sectors of the economy. The park of metal-working equipment in the USSR exceeds in quantity the total park of such equipment in the United States, West Germany and Japan combined... As a result of the dominance of the technology of cutting [as against forging-and-processing] there is 25 per cent more metal used per unit of engineering output than in the United States." (p. 64)

"Inactivity coefficient" - *Pravda,* 12th September, 1982

"The existing grain-drying machine, designed about fifty years ago, cannot cope with today's complicated tasks. Furthermore the capacity of all the grain-driers covers only little more than half of present needs." In a Kazakhstan research centre the engineers designed a grain drier based on new principles, so-called "recirculatory" drier. This was approved by the all-union Academy of Agriculture (VASKhNIL). In 1974, tests were authorised by the Ministry of Agriculture and *Sel-khoztekknika* (which distributes machinery to farms). In 1975 approval was obtained from the Ministry of Tractors and Agricultural Machinery, from Gosplan, and action was to be taken. Another decision was taken in 1978: the machine was "recommended for use in agriculture". But alas, paper decisions are quickly made, but it takes longer to turn them into action. Papers multiplied, but not a single machine was made. Meanwhile the obsolete machines, designed fifty years ago, continue to be produced, together with various attachments... In March 1982 the same official of the all-union Academy of Agriculture wrote again, and again "everyone agreed, no one was against, but still the new grain-driers are not in production".

Pravda, **4th May, 1981** explains the lack of simple grass-cutters (mowers) by the fact that it contributed so little to aggregate plan fulfilment ("mini-technique: mini-interest").

From V.K. Fal'tsman, *EKO No. 7,* 1983

"The share of imported equipment in the installed total tools of labour increased rapidly in the last five year plan period and reached a third."

Comment: About two-thirds of Soviet machinery imports come from Eastern Europe. Therefore the rather high figure quoted by Fal'tsman would indicate that in terms of Soviet domestic prices, the share of Western equipment might be as high as about ten per cent; or, about twice the estimate by Philip Hanson cited earlier.

Appendix 2

A SELECTION OF DATA

The data displayed in Tables 1-4 are from Weitzman (1979), but originally from the Stanford Research Institute – Wharton Econometric Forecasting Associates data bank. The numbers are the same as those employed in Green-Levine (1978), and a fuller description is contained in that source.

Briefly, total capital stock is the official Soviet "basic funds" series in 1955 rubles, adjusted by Cohn. The capital of a given year is the stock in place on 1st January of that year.

Imported Western capital is a constructed series built up out of import data and made compatible with the total capital series. A discussion is contained in the Green-Levine and Green-Jarsulic papers. Weitzman (1979) notes that because of ambiguities inherent in valuation and interpretation, and also because of the unavoidable assumptions involved in converting investment data into capital stock, the imported Western capital series is easily the most controversial of this study. Note also our discussion of data in Part D5.

Domestically produced capital (more accurately, capital of non-Western origin) is defined as the difference between total and imported Western capital.

Labour figures are due to Rapaway at the Department of Commerce, and are measured in thousands of persons employed.

Output indices are the standard Office of Economic Research index numbers, compiled in 1976.

Table 1. **USSR: total industry**

I	K	K_i	L	Y
1960	80	1.39	22 620	53.4
1961	88.7	1.64	23 817	56.25
1962	100	1.92	24 667	60.41
1963	111	2.22	25 442	63.88
1964	124	2.47	26 317	67.36
1965	137	2.65	27 447	72.45
1966	150	2.78	28 514	77.28
1967	163	2.93	29 448	83.41
1968	176	3.21	30 428	88.83
1969	190	3.66	31 159	93.48
1970	208	4.18	31 593	100
1971	227	4.47	32 030	106.24
1972	246	4.62	32 461	111.64
1973	266	5.02	32 875	118.24
1974	288.5	5.42	33 433	125.47
1975	313.3	5.99	34 054	132.76

K is total stock of fixed assets in billions, 1955 rubles.
K_i is the stock of imported capital assets in billions, 1955 rubles.
L is total employment in thousands of persons.
Y is the index of output.
These definitions apply also to Tables 2-4 below.

Table 2. USSR: machine building and metal working

I	K	K_i	L	Y
1960	16.2	0.53	7 206	47.05
1961	17.7	0.6	7 682	49.84
1962	20	0.67	8 189	56.07
1963	22.3	0.75	8 729	59.84
1964	24.4	0.82	9 305	64.36
1965	27	0.91	9 905	69.07
1966	28.5	0.97	10 400	73.47
1967	31.7	1.02	19 846	79.55
1968	34.7	1.08	11 282	86.55
1969	37.4	1.19	11 698	93.16
1970	41	1.35	12 017	100
1971	45.6	1.58	12 369	109.73
1972	50.8	1.69	12 718	119.21
1973	55.7	1.86	13 049	130.65
1974	61.22	2.06	13 424	142.81
1975	67.67	2.31	13 816	154.29

Table 3. USSR: petroleum products

I	K	K_i	L	Y
1960	5.8	0.25	196	40.05
1961	6.3	0.3	200	45.12
1962	6.9	0.37	204	51.22
1963	7.5	0.47	205	57.3
1964	8.3	0.58	221	62.06
1965	9.1	0.72	229	68.24
1966	10.1	0.86	242	74.24
1967	11.1	0.94	252	80.92
1968	12	1.04	254	86.63
1969	13.9	1.17	256	92.29
1970	15	1.28	263	100
1971	16.3	1.36	263	106.91
1972	18.1	1.45	265	114.02
1973	19.67	1.51	268	122.31
1974	21.62	1.54	268	131.78
1975	23.85	1.63	269	141.2

Table 4. USSR: chemicals and petrochemicals

I	K	K_i	L	Y
1960	3.9	0.19	792	38.19
1961	4.6	0.31	868	41.77
1962	5.2	0.41	951	46.47
1963	7.2	0.47	1 042	50.87
1964	9.1	0.55	1 142	57.39
1965	10.7	0.62	1 251	66.12
1966	12.5	0.68	1 346	72.48
1967	13.7	0.76	1 424	79.77
1968	15.7	0.87	1 468	85.07
1969	16.3	0.98	1 523	90.31
1970	18.1	1.04	1 568	100
1971	20.7	1.06	1 598	108.07
1972	22.8	1.09	1 626	115.07
1973	24.5	1.17	1 667	125.08
1974	26.95	1.23	1 706	137.2
1975	29.6	1.29	1 753	152.69

The data displayed in Tables 5, 6 and 7 are from Hanson (1982), and a fuller description is contained in that source.

Table 5. USSR, Poland and Hungary: imports of machinery and equipment (SITC 7) from OECD Countries, 1969-1979
FOB, current prices, $MN

	USSR	Poland	Hungary	All Eur. CMEA
1969	1 060	263	105	2 123
1970	1 028	241	140	2 218
1971	903	283	196	2 300
1972	1 207	577	235	3 112
1973	1 729	1 090	280	4 415
1974	2 309	1 528	408	5 980
1975	4 576	2 084	471	9 351
1976	4 909	1 990	535	9 454
1977	5 375	1 865	725	10 283
1978	5 816	1 857	978	11 408
1979	4 851	1 598	919	11 156

Source: OECD, *Statistics of Foreign Trade*, Series B and C (for 1970-77), as cited in Zaleski and Wienert, 1980, Tables A-6, A-13, A-21, A-24.

Table 6. USSR, Poland and Hungary: imports of machinery and transport equipment (SITC 7) from OECD Countries, 1971-1979
FOB, current prices, $MN

	USSR	Poland	Hungary	All Eur. CMEA
1971	1 436	468	341	3 758
1972	1 759	729	366	4 631
1973	2 164	1 233	357	5 533
1974	2 574	1 729	467	6 693
1975	4 576	2 084	471	9 351
1976	4 785	1 925	516	9 170
1977	4 761	1 643	631	9 068
1978	4 443	1 415	732	8 662
1979	3 400	1 103	614	7 699

Table 7. USSR, Poland and Hungary: percentage shares of imported western machinery in total domestic machinery investment, 1971-1979

	USSR	Poland A	Poland B	Hungary
1971	3.9	–	5.9	23.4
1972	3.0	7.4	8.5	27.4
1973	3.3	12.3	14.8	23.1
1974	3.7	16.2	19.1	23.7
1975	4.0	18.5	22.1	21.8
1976	6.7	20.2	23.7	22.7
1977	6.6	15.3	19.2	26.6
1978	6.0	13.5	16.5	28.6
1979	5.4	11.5	14.7	25.4
1980	–	–	–	22.8

A. B-colum corrected by the ratio of OECD-reported OECD SITC 7 exports to dollar value at official exchange rates of Polish-reported CTN1 imports from non-socialist economies in year t-1.
B. CMEA classification (CTN1) machinery and equipment imports from non-socialist countries as percentage of all such imports in year t, times share of all imported machinery and equipment in total machinery equipment investment in year t+1.

Book II

**TRANSFER OF TECHNOLOGY
FROM WEST TO EAST:
A SURVEY OF SECTORAL CASE STUDIES**

by
George D. Holliday

Book II

TRANSFER OF TECHNOLOGY
FROM WEST TO EAST:
A SURVEY OF SECTORAL CASE STUDIES

by

George D. Holliday

I. INTRODUCTION

The purpose of this study is to survey and assess a specific part of the existing literature – sectoral or branch case studies – on the transfer of technology from Western industrial countries to the Soviet Union and Eastern Europe[1]. The analysis of existing case studies is intended to shed light on two central questions about the technology transfer process – the assimilative or absorptive capacities of the Soviet and East European economies and the impact of West-East transfers of technology on the general level, direction and composition of East-West trade.

The assimilative capacity of the Soviet and East European economies refers to the ability of enterprises in those countries to import, put into operation and diffuse foreign technologies. An understanding of assimilative capacity may contribute to a better assessment of the actual and potential impact of Western technology on the Eastern economies. For example, the contributions of Western transfers of technology to the technological level of Eastern industries, to Eastern military prowess, and to general economic growth in the East are likely to be affected by assimilative capacity.

The questions of assimilative capacity and impact on trade are related. On the one hand, to the extent that Eastern economic planners make rational import decisions, the level of technology imports from the West will be limited by the speed and efficiency with which they can be assimilated. Thus, if Eastern importers cannot put imported machinery and equipment to productive use, they are likely to turn to domestic alternatives, reducing the level of imports from the West. On the other hand, Eastern planners have pursued a strategy of using imported technology to build export-oriented production facilities that are competitive on Western markets. The ability of the Eastern economies to expand exports depends partially on assimilative capacity.

There is a growing literature on Western technology transfer to the Soviet Union and Eastern Europe, much of it relating to political and strategic questions, corporate strategy, or macro-economic aspects of technology transfer. In addition, there are a number of general surveys of the actual experiences of Western firms involved in West-East technology transfer which provide valuable insights into the questions of assimilation and trade, but do not differentiate on a case-by-case basis. The focus of this survey is on those studies which have looked at the West-East technology transfer process by sector or by project. An intended result is to complement the more general assessments of technology transfer by comparing and contrasting the experiences in various discrete technology transfers to determine if there are significant differences in assimilative capacity or trade impact. Are some Eastern projects more likely than others to assimilate Western technology rapidly and efficiently? Are some more likely to generate exports which compete on Western markets? Are some industrial sectors in the East more likely to attract additional imports of Western technology and intermediate goods in the future?

Political and strategic aspects of technology transfer to the East have been central concerns of Western policymakers. To what extent do various Western technologies contribute to Eastern military capabilities? Does Eastern dependence on Western technology provide Western governments with significant leverage in East-West diplomacy? What are the political consequences of growing Eastern indebtedness to Western creditors? Will increased competition from Eastern exporters cause economic dislocations and adverse political reactions in the West? Generally speaking, these aspects of Western technology transfer to the East are not analysed directly and in detail in the case studies included in this survey. Nevertheless, the findings of the case studies may have significant political and strategic implications because the answers to such questions depend in large part on assumptions about the assimilative capacities of the Eastern economies and the trade impact of West-East technology transfers.

1. ASSIMILATIVE CAPACITY

In analysing the assimilative capacity of the Eastern economies, the primary concerns are the speed and efficiency with which foreign technologies are put into operation. With respect to the speed of assimilation, one can usefully distinguish four aspects[2]:

    a) the initial lead-times for importing technology;
    b) improvements in lead-times for successive imports of technology;
    c) the extent to which the technology is diffused throughout the economy; and
    d) the speed of diffusion.

The initial lead-time is the time taken to acquire, install and begin operating a new technology the first time it is imported. Subsequent imports of the same technology can be expected to be more rapid as the technological capabilities of the recipient improve. Diffusion refers to the process by which the imported technology is incorporated throughout the industry or the economy.

The initial lead-time is generally the most important determinant of the technological level of a leading plant or the most advanced part of an importing industry. The improvement of lead-times and the extent and speed of diffusion, however, are more important determinants of the general technological advancement and growth of the industry.

The efficiency of assimilation refers to the degree to which foreign technoiogy is put into operation with an optimal level of output, a satisfactory level of quality, and a minimum cost. Thus, the efficiency of assimilation relates to the costs – expenditures on both imported and domestic inputs – of putting a new technology into operation and to the achievement of output goals. Assimilation may be speedy, but not efficient, if resources are wasted in the process and the end result is less than optimal output.

Assumptions about the assimilative capacities of the Eastern economies underly much of the Western literature about technology transfer to the East. The assumptions, however, have not always been made explicit and have rarely been tested empirically. Generally, it has been assumed that the Eastern economies have assimilated technology relatively slowly and inefficiently. Many examples of slow adaptation, diffusion and updating of Western technologies have been cited in the literature. The existence and persistence of technology gaps between Western industrial sectors and their Eastern counterparts are sometimes noted as evidence of slow assimilation.

The assumption of slow assimilation, however, is by no means universally shared among Western observers. Some Western assessments of the contribution of Western technology to economic growth in the East, for example, assume very rapid and efficient assimilation of Western technology. For example, several econometric studies attempting to measure the impact of Western technology on Soviet economic growth have been based on assumptions of rapid assimilation [82][3].

Assessments of the assimilative capacities of the Eastern economies have clear implications for some of the political and strategic issues identified above. For example, an assessment of the assimilative capacity of the Soviet Union is a prerequisite to understanding the contribution of Western technology to Soviet industrial growth and military potential. Thus, if Soviet industry is found generally to assimilate Western technology slowly and inefficiently, policymakers would presumably be less concerned about the strategic implications of technology transfers. On the other hand, if certain types of contractual arrangements with Western firms are found to facilitate effective assimilation or if certain Eastern sectors are found to have good assimilative capacities, Western export control authorities might be more concerned about the strategic implications. In short, Western export control policies may be guided to some extent by a better understanding of the conditions under which assimilation is likely to be rapid and efficient.

2. IMPACT ON TRADE

The transfer of technology from West to East influences trade flows in both directions: Western shipments of machinery, equipment and industrial plants, with accompanying technical assistance to the East, sometimes assist Eastern shipments of resultant products to the West. In each of the Eastern countries, increasing expenditures on imports of technology from the West have necessitated a greater emphasis on expanding exports to the West. A major rationale for importing Western technology is the creation of modern industries which produce goods that are competitive on Western markets. Indeed, technology imports from the West are frequently tied directly to exports by means of various contractual provisions for counterpurchases or product buybacks. Even when no such contractual provision exists, Eastern managers frequently allocate a part of the output of Western-assisted projects for export to Western markets.

Commercial technologies are most often transferred either embodied in modern machinery and equipment or in the form of accompanying technical and managerial assistance, including licenses, technical specifications, training, startup assistance and servicing. (Since complementary technical assistance is frequently included in the price of machinery and equipment exports, total technology transfers are often measured, albeit imperfectly, as the value of machinery and equipment exports.) Thus, the transfer of technology from West to East stimulates exports of capital goods from Western countries, providing the usual gains from international trade – increases in income and employment and economies of scale in the affected industries – and balance of payments benefits.

Trade statistics compiled by Western observers show clearly which Western machinery manufacturers have benefited most from West-East technology transfer. Those who produce machinery and equipment for the chemical, automotive, oil and gas, metalworking and metallurgical, electronics, shipping, mining and construction sectors, for example, were consistent leaders in the transfer of technology to the Soviet Union in the late 1970s (See Table).

The smaller East European countries have shown a broadly similar pattern of machinery and equipment imports, concentrating on the chemical and engineering industries. There have been, however, a few notable differences. Not surprisingly, they

USSR: Machinery orders placed with hard currency countries
Million US $

	1976	1977	1978	1979*
Total	5 991	3 816	2 803	2 612
Of which:				
Chemical and petrochemical	1 818	1 628	902	607
Oil and natural gas	1 688	308	832	190
Metal working and metallurgy	1 028	641	348	752
Timber and wood	146	65	86	56
Automotive	355	183	115	184
Ships and port equipment	283	67	127	61
Food processing	63	155	17	24
Mining and construction	120	147	118	149
Manufacturing of consumer goods	121	78	44	12
Electronics	55	193	179	335
Electricity	63	138	6	30

* Estimated.
Source: US Central Intelligence Agency, National Foreign Assessment Center, *The Soviet Economy in 1978-79 and Prospects for 1980*, ER-80-10328, June 1980, p. 17.

have imported less technology for oil and gas industries. Such light industries as textiles and clothing, on the other hand, have received relatively more emphasis in the smaller East European countries [79].

While trade statistics provide a useful overall picture of West-East transfers of technology, they cannot provide conclusive answers to some important questions concerning the impact of technology transfer on trade. For example, do transfers of technology free the East from future reliance on imports from the West, or, conversely, do they generate demand for complementary technologies and intermediate goods and for periodic technological updates? Because changes in the level and composition of trade have multiple causes, trade data alone provide somewhat ambiguous answers to such questions.

Assessing the impact of technology transfers on Eastern exports to the West is also problematical. While trade statistics show the commodity composition of Eastern exports to the West, they do not show the relationship between imports of technology and resultant exports. Some general studies of technology transfer have tended to discount the ability of Eastern industries to generate exports that are competitive on Western markets. Drabek, for example, finds little statistical evidence (at the broad sectoral level) that Western technology has contributed significantly to East European exports to the West [79]. He acknowledges, however, that at a more disaggregated level, statistics may reveal evidence of Western technology contributing to Eastern exports in some subsectors. Parts of the Eastern chemical industries may provide examples. Trade policymakers in the United States and Western Europe have already been confronted with complaints about market disruption from Eastern exports produced with the assistance of Western technology.

The answers to such questions are important to students of the Eastern economies, to Western policymakers concerned with trade and strategic issues and to Western commercial interests. In the late 1960s and 1970s, trade with the West began to play an increasingly important role in the economic growth strategies of the Soviet Union and Eastern Europe. During that period, the Eastern countries as a group suffered chronic deficits in their hard currency trade accounts. As a result, the six East European countries and the Soviet Union accumulated a hard currency debt estimated at approximately $80 billion in 1981. For some of the countries, such as Poland, Hungary and Roumania, servicing the debts to the West became a significant economic burden; for all, shortages of hard currency became a major constraint on their ability to trade with the West.

While Western creditors, both private and official, may continue to extend credits to the East, they are unlikely to expand their exposure there as rapidly as in the 1970s. Ultimately, the Eastern countries can continue to import Western technology only to the extent that they succeed in exporting to the West. Thus, the degree to which Eastern economic planners are successful in building modern industries capable of competing in Western markets has profound implications for future East-West trade. Eastern export competitiveness is an important consideration for Western firms seeking export markets in the East and for Western firms which might be confronted on Western markets by Eastern competitors. Eastern competitiveness is a crucial consideration for Western creditors (including official lenders and guarantors) who must make decisions on credit worthiness. It is also an important consideration for Western policymakers concerned with strategic issues: in the long run, Eastern debt problems could be a more significant constraint than Western export controls on West-East technology transfers.

Case studies are unlikely to provide comprehensive answers to the questions concerning assimilation and trade. They could, however, shed light on the conditions under which Western technology is likely to be assimilated rapidly and efficiently and have a significant effect on East-West trade. This survey examines existing case studies for insights into the questions of assimilation and trade. It also attempts to identify gaps in the literature and poses specific questions which future case studies might address in greater detail.

II. CASE STUDIES

In selecting case studies to be included in the survey, several criteria were applied in order to focus the survey clearly on the questions of assimilative capacity and trade impact. Only major case studies with substantial information and analysis of the questions related to assimilative capacity and trade impact were selected. While many general studies contain useful information related to individual cases or examples of technology transfer, they were omitted in order to narrow the scope of the survey to manageable proportions. Particularly noteworthy among the general studies are a number of surveys, based on interviews and questionnaires, of the actual experiences of Western firms involved in West-East technology transfer. They provide valuable insights into the questions of assimilation and trade, but do not differentiate on a case-by-case basis. (See especially ECE [92], Bolz-Plötz [78], McMillan [86], and Marer-Miller [87].) Since they are ably summarised and discussed in Levcik-Stankovsky [41] and Zaleski-Wienert [96], they are merely referenced and not discussed in detail in this survey.

The survey is also limited to case studies of actual technology transfers, most of which took place in the 1960s and 1970s. Omitted are feasibility studies and market surveys carried out for Western companies considering entry into the Eastern market. (Such studies sometimes provide a wealth of detail about specific projects and statistical data on Eastern industrial sectors. See for example, Chase World Information [8] on the Kama River Truck Plant and Internationales Zentrum-Chase World [34] on the East European automobile industries.) Also omitted are studies of prospective projects which have not actually taken place. (For example, see Kosnik [38] for a discussion of prospective Western-assisted natural gas projects in the Soviet Union.) The survey is limited mostly to case studies of transfer of technology since the early 1960s simply because most case studies have dealt with that period, and the experience of that period is likely to be most relevant to policy questions related to assimilation and trade.

The survey uses a common framework for summarising the case studies in order to facilitate comparison and analysis. In order to collect and organise information related to assimilative capacity and trade impact, the survey examines each of the case studies for analysis of the following questions:

a) *Methodologies and sources*

What approaches to studying individual transfers of technology have been used?

What are the primary sources of information and data?

b) *Level of technology and degree of dependence on the West*

Do Eastern importers obtain the latest available Western technologies? What is the contribution of Western technology to various sectors of the Eastern economies? To what extent are they dependent on the West for technological progress?

Do the contracts provide for one-time transfers of technology or for continuing updates of technology?

c) *Channels or mechanisms of technology transfer*

What types of contractual arrangements have been used by Eastern importers of Western technology?

How does the choice of a mechanism for technology transfer affect the assimilation process?

d) *Hard currency constraints and efforts to expand exports*

What are the total foreign exchange expenditures in technology transfer transactions?

How do Eastern foreign trade officials attempt to overcome hard currency constraints?

To what extent are technology purchases tied to future Eastern exports through buy-back or counter purchase arrangements?

How successful have Eastern exporters been in competing on Western markets?

e) *Indicators of assimilative capacity*

How do lead-times for imported plants in the East compare with similar plants in the West?

How efficiently do imported technologies operate after startup?

What is the speed and extent of diffusion to other parts of the economy?

f) *Effects of domestic infrastructure and institutions on assimilative capacity*

What level of domestic expenditures are needed to complement technology imports?

How do weaknesses and strengths in the domestic economic and technological infrastructures of the Eastern countries influence their assimilative capacities?

What is the quality of domestic work forces, supplier industries, research and development facilities, and other domestic inputs?

How is the assimilative capacity of an industry affected by the domestic economic and political system?

The approaches and findings of major case studies are summarised and compared in this chapter. The final chapter provides general comments on the findings of the case studies and identifies gaps and weaknesses in the existing literature.

1. METHODOLOGIES AND SOURCES

With the rapid expansion of Western technology transfer to the East, the number of Western corporate executives who have participated in and observed first hand the technology transfer process has grown considerably. A number of researchers have found such corporate executives to be a valuable source of information for case studies. Indeed, most case studies have made at least some use of this source of information although some have merely casually collected anecdotal information and used it to supplement other sources. A few authors of case studies, on the other hand, have surveyed Western corporate executives systematically, employing standardised questionnaires in an attempt to acquire comparable information and data from a cross section of Western firms involved in the transfer of technology to the East.

Prominent among the more systematic surveys are two studies commissioned by the Stanford Research Institute, one by Hanson and Hill [27] of the experiences of United Kingdom firms involved in transferring technology to the Soviet Union, and another by Röthlingshöfer and Vogel [51] of German firms' involvement in transferring technology to the Soviet Union. The Hanson-Hill study analyses information provided by eight companies in the machine-tool sector and seven companies in the chemical equipment sector, most of which have been involved in numerous projects in the Soviet Union. The Röthlingshöfer-Vogel study analyses responses from executives of 14 companies, most engaged in plant construction in the metallurgical, chemical, food and packaging industries.

The two studies are among the few that focus squarely on the question of assimilative capacity of Soviet industry. More precisely, the Hanson-Hill study focuses on lead-times, the times taken to acquire, install and begin operating imported machinery. A secondary purpose of their study was, where possible, to compile data on the output and manning levels of new plants when they began operating in the Soviet Union and on the rate of diffusion of technologies embodied in imported machinery to domestically-built machinery. Similarly, the purpose of the Röthlingshöfer-Vogel study was to investigate Soviet reasons for importing machinery and equipment from the Federal Republic of Germany, and to analyse the course of and time scales for such transactions. They also sought to determine whether such imports have been important in raising output in the importing industry.

The Hanson-Hill and Röthlingshöfer-Vogel surveys employed the same survey techniques and questionnaires. The authors sent letters requesting information from firms that were known to be actively involved in transferring technology to the Soviet Union. Those corporate executives who gave positive responses received a questionnaire which provided a standardised structure for discussion at subsequent interviews. Information received at the interviews was then written up in the form of a series of case studies. (In the case of the British chemical equipment firms, the responses were summarised in an aggregative form in order to avoid divulging proprietary information.) The questionnaires used in the two surveys focused on several kinds of information.

- a) Technical background. (For example, they asked questions about the types and level of technology being transferred, the role of Western and Soviet technicians in the transfer process, and the degree of Soviet technological lag.)
- b) Proposal and contract. (The questions dealt primarily with the time required for this stage.)
- c) Acceptance and installation. (The questions focused on the time required for this stage and the reasons for delays, if any occurred.)

d) Total time from initial inquiry to installation. (They asked respondents to make comparisons with West European customers and explain reasons for differences.)

e) Utilisation. (Questions dealt with the effectiveness with which the technology was used and with manning levels.)

f) Diffusion. (They asked questions about the subsequent incorporation of the imported technology in the design of Soviet-built machinery and the time required for this process.)

A series of studies carried out by the United Nations Economic Commission for Europe (ECE) also relied heavily (though not exclusively) on interviews with corporate executives. For example, a case study of East-West industrial co-operation in the machine-tool sector was based on interviews with executives from 15 Western firms which had been involved in such agreements with Hungary, the USSR, Poland, Roumania and Czechoslovakia [63]. The purpose of the study was to describe the forms of co-operation that had been undertaken, the motives of the participants and the problems encountered. Questions dealt with five general areas: the forms and modalities of transfer; the motives for entering co-operation agreements; preparation of the agreement; implementation of the agreement; and overall assessment of the agreement. Another ECE case study, dealing with East-West co-operation in the automotive sector [61] also used information from interviews with corporate executives, but supplemented it with much information from published sources, especially Western and Eastern journals specialising in the automotive industry and East-West trade. Both studies presented summaries of the acquired information instead of case studies of specific transactions.

Among the other case studies based at least in part on interviews are those by Hayden of eight United States firms involved in the transfer of technology to Poland and Roumania [30], McMillan-St. Charles of joint ventures in Roumania, Yugoslavia and Hungary [43], Baranson of technology transfers by United States firms to foreign enterprises located in both Western and Eastern countries [1], Centre d'études industrielles (CEI) of Fiat operations in Poland [7], Garland-Marer of industrial co-operation between International Harvester and the Polish foreign trade organisation BUMAR [19], and De Pauw of Control Data Corporation's negotiations with Soviet trade officials [12]. The Garland-Marer study appears to be unique among the case studies surveyed in that the authors received extensive information about individual transactions from managers and engineers of both Western and Eastern partners. While several other authors of case studies allude to interviews with Eastern country officials, none seems to have received the kind of co-operation from Eastern managers and technicians at the firm level reflected in the Garland-Marer study. De Pauw's study is noteworthy for the quality of information obtained from Western corporate sources. He used standardised questionnaires, interviews with corporate officials and corporate documents to study the experiences of United States firms in negotiations with Soviet officials.

Authors of case studies have also made extensive use of Western periodical literature. In particular trade journals specialising in East-West commercial relations or in various industrial sectors provide information based on interviews with Western corporate executives, government officials and others. While the information and data from such sources are frequently superficial, they occasionally provide useful insights into questions of assimilation and trade. Several examples may illustrate the utility of Western periodical literature for this purpose. The newsletters *Business Eastern Europe* and *East-West* routinely publish information on the problems of negotiating and implementing East-West commercial transactions. *Soviet Business and Trade* is a useful source of information on new contracts negotiated with the Soviet Union. The Economic

Intelligence Unit's *Quarterly Economic Review of the USSR* (and its reviews for other Eastern countries) provide valuable data about the volume of machinery orders for major industries. The French journal *Le courrier des pays de l'Est* publishes articles on East-West co-operation and trade in various industrial sectors. These sources have been cited frequently by authors of case studies. Hanson-Hill and Sobeslavsky-Beazley [55] cite the journals *Chemical Age, European Chemical News,* and *Chemical Insight* as useful sources for data on East-West contracts in the chemical sector. An article in *Modern Casting* by Schaum [52] provides unusual detail of assimilation problems at the Kama River Truck Plant. For other sectors, the authors of various case studies have cited articles in such technical journals as *Oil and Gas Journal, Power Engineering* and *Automotive Industries.*

The authors of a number of case studies have also made extensive use of published sources in the Eastern countries. In particular, studies published by the US Central Intelligence Agency, National Foreign Assessment Center on the Soviet automobile [68], chemical [67], and studies by Tasky on the Soviet and East European computer industries [59, 60], a study by Holliday on transfer of technology to the Soviet automotive industry [32], the ECE study on industrial co-operation and trade in the automotive sector, and a study by the US Office of Technology Assessment (OTA) on the Soviet energy sector [71] appear to be based in large part on information and data from Eastern sources. Eastern periodicals sometimes provide especially useful information on the rationale for importing technology and problems of assimilation. The technical and economic literature in Poland and Hungary appear to be particularly useful sources of information. Some Soviet periodicals are also useful. On the other hand, sources from Czechoslovakia, the German Democratic Republic and Roumania have apparently provided relatively little useful information.

Hanson makes extensive use of Western and Eastern published sources in an attempt to measure the impact of Western technology on the Soviet mineral fertilizer industry [25]. Because Soviet foreign trade statistics do not provide adequate, disaggregated data for chemical equipment imports, he relies on Western press reports and industry sources to compile a detailed picture of Western plant contracts. To identify and quantify the role of imported plant, Hanson uses the following procedures:

- a) he compares total known capacities of imported plant during 1960-1975 with growth in Soviet capacity or output of the relevant product during the same period;
- b) he compiles Soviet plant start-up data, derived from Soviet radio broadcast sources (for which he uses the *BBC Summary of World Broadcasts,* Weekly Economic Report series); and
- c) he obtains supplementary information from Soviet press articles and specialist literature.

With these data he estimates how much of each major kind of fertilizer has been produced with imported plant, the net contribution to agricultural output of the fertilizer produced with imported plant, and the extent of domestic diffusion of commercially imported technology.

In its study of Western technology transfer to Soviet energy industries, OTA [71] combines surveys of the Soviet oil, gas, coal, nuclear and electricity industries with Western trade data. For the industry surveys, OTA relies heavily on the Soviet technical and economic literature. The surveys include a largely qualitative discussion of the contribution of Western technology to each industry. Its primary sources for trade data are the United Nations' Standard International Trade Classification (SITC), United States Department of Commerce trade statistics (Schedules B and E), the European Economic Community's Nomenclature of Goods for the External Trade Statistics of the

Community and Statistics of Trade Between Member States (NIMEXE), and the bi-weekly publication *Soviet Business and Trade*. From these sources, it attempts to quantify the transfer of technology from the West to the Soviet energy sector.

2. LEVEL OF TECHNOLOGY AND DEGREE OF DEPENDENCE ON WEST

Most of the case studies address several key questions related to the level of technology transferred and the degree of Eastern dependence on Western technology. There is substantial agreement among the authors on the broad outlines of the answers to these questions. Most case studies conclude that Eastern purchasers have sought recent (though not always "the latest") technology from the West. There is general agreement that at least one important reason for Eastern purchases is retardation of domestic technologies and reliance on the West to spur domestic technological progress and economic growth. Finally, most authors suggest some degree of dependence on Western technology continuing after completion of the transfers which they describe. Within these broad generalisations, however, there are some significant differences of emphasis.

a) *Level of technology transferred*

The authors of the case studies generally agree that Eastern purchasers want the most advanced Western technologies available. Their choice of Western technologies is almost always complicated, however, by the desire to select proven technologies or by resistance from Western firms, which prefer not to sell their latest technologies. For example, the Röthlingsöfer-Vogel case studies conclude that Soviet buyers look for the most advanced technologies, but insist on evidence that the technologies are commercially proven. In practice, this frequently means that Soviet specialists must see the technology operating successfully in the West. Given the long lead-times required to put the technology into operation in the Soviet Union, it may be somewhat dated by the time the technology transfer is completed. Three of Hayden's case studies – the Honeywell, Inc., and Singer Co. agreements with Poland and the Control Data Corp. agreement with Roumania – provide examples of Western companies declining to sell their latest technologies, preferring to sell older technologies while exploiting the latest technologies for the companies' own commercial advantage.

Radice [50] discusses another reason why the latest technology may not be transferred. The technological retardation of the electronics industries in Eastern Europe, he maintains, makes it difficult to assimilate the latest technologies from the West. He suggests that it is impossible to jump to the most modern products and production processes without first building up a base of skill and experience.

Western export control laws may also be a factor in restricting transfers of the latest technologies. Tasky [59] notes that the relaxation of US export controls was followed by an upsurge in sales of computers to Eastern Europe, and Grant [22] observes a similar pattern in sales of numerically controlled machine-tools to the Soviet Union. Such a pattern suggests that export control laws and regulations may limit Eastern purchases to older technologies in some sectors. The most sophisticated Western computers and machine-tools, for example, are still not available to Eastern buyers.

Several case studies provide examples of efforts by Eastern importers to establish long-term relationships with Western firms that ensure continual updates of the imported technologies. Marer's composite case study of United States chemical equipment suppliers [45] concludes that in order to become a principal contractor in an Eastern European country, the supplier must generally be able to offer process and formula licenses that are among the most modern anywhere. According to Marer, they must also be willing to pass on improvements for approximately five years after startup. The ECE study of East co-operation in the automotive sector [61] describes a major change in co-operation agreements: while those signed before 1975 were based on existing product designs already in production, agreements signed after 1975 tended to include commitments on the part of Western firms to provide new vehicle designs not yet available on Western markets and advanced production techniques and equipment with which to produce a major portion of the vehicle's components. Gutman describes a similar pattern of co-operation in the automotive industry [24].

Baranson also notes cases in which United States firms have sold "their most recently developed technology", and emphasizes that the new kinds of technology transfer agreements are enabling a rapid and efficient transfer of those technologies [1]. Baranson also describes a case – an agreement between General Motors and Polmot – in which the partners agree to co-operate in the design and engineering of a new or improved product line. Garland-Marer's study of the International Harvester-BUMAR agreement [19] notes that the Western firm is obligated to continually provide its partner with its most current technology. Holliday describes the long-term, but somewhat looser, ties between Fiat and the Soviet automotive industry – a scientific and technological co-operation agreement – and suggests that it is a likely prototype for future Soviet efforts to maintain long-term technological ties with Western firms. Grant concludes that the Soviet machine-tool industry is relying on similar agreements (he lists 16) for design and manufacturing technology needed to produce modern machine-tools. These and other authors suggest a perceived need on the part of Eastern economic planners for large-scale and long-term technological ties to the West in order to keep abreast of technological progress.

b) *Eastern dependence on Western technology*

Perhaps the greatest differences exist regarding the generalisation about Eastern dependence on Western technology in various industrial sectors. At one extreme is Sutton's "inability thesis" about the Soviet Union: he describes cases in numerous sectors of Soviet acquisition of Western technologies and concludes that there has been virtually no indigenous Soviet innovation [58]. He maintains that Western technological assistance has been the major causal factor in Soviet economic growth. Sutton's extreme conclusion is not supported by several efforts to quantify the contribution of Western technology to the Soviet economy [82]. Nor does it conform to the findings of authors of most other case studies, although all tend to attribute to Western technology a significant role in the Eastern economies. (Support for Sutton's conclusion is found in a study by Judy of the Soviet computer industry [35]. In the late 1960s, Judy concluded that Soviet computer technology was virtually entirely imported from the West.)

Most other authors tend to emphasize that there is some degree of dependence on Western technology for modernisation of key industrial sectors in the East. Thus, the perception that domestic industries are technologically obsolescent compared to their Western counterparts is considered to be a major motivating factor in Eastern purchases of Western machinery and equipment. Typically, Röthlingshöfer-Vogel [51] conclude on the basis of their survey of German suppliers that Soviet purchases were chiefly initiated because of the general technological backwardness of Soviet industry, a technology gap which they suggest is perhaps ten years.

Technological dependence is not the only reason for Eastern purchases of Western technology, however. Röthlingshöfer-Vogel conclude that shortcomings in Soviet production capacity were a second important reason. Finally, in some cases, it was not efficient to develop domestically what was already available abroad. Similarly, the Hanson-Hill survey of United Kingdom exporters of machine-tools [27] emphasizes both economic and technological factors. They conclude that the purchases were initiated because of shortcomings in Soviet production capacity. However, once the decision to import was made, they sought technologies that were superior to comparable Soviet-produced items. Sobeslavsky-Beazley, while emphasizing Soviet dependence on imports of Western chemical plants point out another reason for importing Western chemical plants: Western firms are more likely to deliver the plants in a more timely fashion than Soviet industry [55].

The most systematic effort to measure the impact of Western technology on an individual sector is Hanson's study of the Soviet mineral fertilizer industry [25]. On the basis of some admittedly crude calculations, he estimates that in 1975 the percentages of Soviet fertilizer output that came from imported Western plants were: nitrogenous, 40; phosphate, 5; potash, 5; and complex, 65. He further estimates that in the mid-1970s about 3.25 per cent of the value of net agricultural output (or about 4.5 per cent of gross crop output at 1965 prices) could be attributed to the application of fertilizers produced in imported Western plants. Hanson uses these estimates cautiously, but concludes generally that the total economic impact of Western technology on the Soviet mineral fertilizer industry and on Soviet agricultural output has been considerable.

While authors of other case studies have not attempted the kind of estimates offered by Hanson, most agree with his general conclusion. A CIA study [67], for example, values Soviet purchases of chemical equipment from the West during 1961-1975 at $5 billion. During 1971-1975, imports were estimated to be from one-fifth to one-fourth of total Soviet investment in chemical equipment (including more than 40 per cent of the equipment used for production, storage and handling of fertilizers, 20 per cent of the equipment for man-made fibres, and 14 per cent for plastics). Similarly, CIA studies of the Soviet oil [65, 66] and gas [64] industries, Tasky's study of the East European [59] and Soviet [60] computer industries, and Geze's study of Eastern electronics industries [20] suggest major contributions of Western technology.

A major work, edited by Amann, Cooper and Davies, on the technological level of Soviet industry [76] does not examine systematically the contribution of Western technology to the various sectors studied. Nevertheless, many of the contributors to the volume addressed the question, and their findings are summarised by Davies. Davies divides the industries studied into three groups. One group, with a strong indigenous technology, has relied relatively little on imports of machinery and equipment. Examples are space rocketry, weapons, nuclear power, and electric power. A second group has a well established indigenous technology and development capability, and has been responsible for pioneering important innovations, but has nevertheless relied on foreign technology in certain respects. The iron and steel and machine-tool industries are cited as examples. For the third group, dependence on foreign technology is very high. Davies cites the Soviet computer and chemical industries as examples. The Amann, Cooper and Davies study and a subsequent study by the same group [77] emphasize that patterns of technological innovation vary widely among Soviet industries. Accordingly, Soviet industries differ in their degree of dependence on Western technology as a vehicle for modernisation.

Similarly, the OTA study of Western technology transfer to the Soviet energy sector finds a differential contribution [71]. It finds a major contribution of Western technology to the Soviet oil and gas industries. Moreover, for technologies related to the construction and operation of gas pipelines, such as large diameter pipe and other equipment (compressor stations and pipelaying equipment) it finds the Western contribution to be "massive". For other energy industries – coal, nuclear, and electric power – however, Soviet dependence on Western technology has been slight. OTA finds that Soviet purchases have been spotty and appear to have been intended to compensate for specific deficiencies in the quantity or quality of domestic equipment, rather than to acquire new technological know-how.

Hanson comes to similar conclusions by comparing Soviet imports of Western machinery with total investment in machinery for various sectors [26]. His calculations show high dependence for the automobile and chemical industries and substantial dependence for the shipping industry in the late 1950s and early 1960s and the oil and gas industries in the 1970s. He finds that technology imports from the West have been important in Soviet computer usage in all economic branches. For the textile and clothing and timber, paper and pulp industries, he finds that imports of Western technology have been of above average importance. In contrast, he finds the contribution of Western technology has been very low for agriculture, electric power, food-processing, the building materials industry, mining, the mechanical engineering industry excluding the automobile industry, and the non-ferrous metals industry. Dependence was also low for the iron and steel industry until the 1970s.

Most of the case studies suggest continued dependence on Western technology for key Eastern sectors in the future, and some of them indicate specific Eastern technological needs that are likely to be filled by imports from the West. A CIA study of the Soviet natural gas industry [64], for example, details continuing Soviet import needs for high-quality, large-diameter pipe and pipeline-related equipment. It suggests that, if the pace of development of gas deposits in Tyumen and the Arctic seas is increased, there may be a greater need for imports in the 1980s. The OTA study of Soviet energy supports this appraisal [71]. OTA finds that the Soviet gas industry's dependence on imports of Western technology will probably increase during the 1980s.

Another CIA study, of the Soviet oil industry [65], maintains that imports of Western technology and equipment are becoming increasingly necessary for the industry's growth. It predicts that, for the foreseeable future, the USSR will have to rely on the West for much of the equipment and know-how to realise its oil production potential, especially as exploration and development requires deeper drilling or takes place offshore, in Eastern Siberia or in the Arctic regions. The OTA study also concludes that the Soviet Union will continue to import petroleum equipment, but its assessment suggests somewhat less Soviet dependence on Western oil technology in the future. OTA finds that the Soviet Union produces large quantities of most of the items it needs for the oil industry, but continues to import because it has difficulty producing sufficient quantities of high-quality equipment. OTA also finds that in three areas – computers, associated software, and integrated equipment sets for oil exploration; offshore technology; and high capacity submersible pumps – Western technology can continue to make a significant contribution.

Tasky's study of the East European computer industries [59] concludes that imports of Western computer technology (but not computers) is likely to continue to grow over the next few years. Similarly, Mundie-Goodman conclude that there is no foreseeable end to the East European need for imports of Western computer technology.

Case studies of the Eastern chemical industries also suggest continued dependence on Western technology. The Hanson-Hill survey of United Kingdom suppliers of chemical equipment to the Soviet Union [27] concludes that there is a general tendency on the part of Soviet purchasers to go back to Western suppliers for repeat contracts on the next generation of technology. Sobeslavsky-Beazley also find that Soviet dependence on imports of Western chemical technology is likely to continue for the foreseeable future [55]. They find that the processes and equipment involved in modern chemical plants are too complex to be mastered through "reverse engineering", and require parts and materials which cannot be produced by Soviet industry.

A notable exception to the general perception of continued technological dependence is a CIA study of the Soviet automotive industry [68]. The study concludes that the Soviet automotive industry has made powerful strides toward self-sufficiency and has the experience, knowledge and resources to stay abreast of world automotive state of the art through its own efforts. The study maintains that the USSR prefers to manufacture most of its own production equipment and is vigorously enhancing its ability to do so. If the USSR has turned to the West in recent times, the study finds, it is mainly for economic and not technological reasons. That is, the domestic machine-tool industry has been unable to produce specialised machinery fast enough for the rapid expansion of the industry in the 1960s and 1970s. Foreign technical assistance has been sought, the study concludes, only for projects of extraordinary size and urgency.

3. CHANNELS OR MECHANISMS OF TECHNOLOGY TRANSFER

What types of contractual arrangements have been used by Eastern importers of Western technology? How does the choice of a mechanism for technology transfer affect the assimilative capacity of the importer? Such questions have been a major focus of the existing case studies.

a) *Evolution of technology transfer mechanisms*

While technology is transferred from West to East through a great variety of channels or mechanisms[4], the case studies devote major attention to what Baranson calls "a new generation of technology transfer agreements" [1]. This concept, also referred to as "sustained enterprise-to-enterprise relationships" by Hayden [30] and "active technology transfer mechanisms" by Holliday [32], refers to a variety of industrial co-operation agreements or other contractual arrangements that transcend simple, one-time sales of products with relatively little technical assistance. Generally, such arrangements involve frequent and specific communications between the Western and Eastern partners, the transfer of proprietary or restricted information, and extensive interchange of managers, engineers and technicians. Various authors discuss the evolution and proliferation of the new kinds of agreements in the 1960s and 1970s, and most suggest a continuation of recent trends.

Several authors suggest that technology transfer mechanisms have been shaped to varying degrees in the Eastern countries by efforts to maintain control over the technology transfer process, to maintain the independence of domestic economies and to insulate local populations from alien ideas. For example, Holliday describes a "central dilemma" of Soviet leaders since the inception of the Soviet state: the balancing of the political costs of co-operating with foreigners with the perceived economic benefits of assimilating Western technology. He suggests that, during periods of intensive Soviet interest in borrowing Western technology, there has been considerable flexibility in adapting foreign trade institutions to the needs of an effective technology transfer process. Thus, he describes an evolution toward more active technology transfer mechanisms since the 1960s, involving long-term agreements and extensive personal contacts between Soviet and Western specialists during all stages of the technology transfer process. While acknowledging the limitations of this evolution (no foreign direct investment by Western companies, limited participation of Western managers, and continued excessive isolation of Soviet technical specialists), Holliday concludes from his case study of the automotive industry that Soviet leaders are showing a greater willingness to experiment with new technology transfer mechanisms.

The study by McMillan and St. Charles [43] demonstrates a more rapid evolution of institutional arrangements in some of the smaller East European countries. The authors investigate changes in the legal, institutional and political environments in three countries – Roumania, Hungary and Yugoslavia – which have permitted establishment of joint ventures with Western firms. (They study four joint ventures – Control Data Corporation's agreement with Roumania to produce computer peripheral equipment; Massey-Ferguson Ltd.'s agreement with Roumania to produce wheel loaders; Semperit's agreement with Yugoslavia for five chemical plants; and Bowmar Canada Ltd.'s agreement with Hungary to produce calculators.) They cite evidence that the joint venture phenomenon is gaining momentum and suggest that it may spread further in Eastern Europe. McMillan-St. Charles maintain that joint ventures provide special advantages to both partners: they offer the best assurance to Eastern partners that the Western associate will provide production technology, managerial expertise and marketing assistance on a continuing basis, while providing Western partners a greater degree of control over operations in the East. (In retrospect, their appraisal of the prospects for wider use of joint ventures appears overly optimistic. Levcik-Stankovsky [41] and Schnitzer [53] discuss some of the impediments to further development of joint ventures.)

Hayden [30] also concludes that joint ventures are a viable option for Western firms seeking co-operative arrangements with Eastern firms. He emphasizes, however, that equity ownership is not a necessary condition for successfully meeting the key goals of both partners. Indeed, his case studies lead to the general conclusion that prospects for other kinds of sustained enterprise-to-enterprise relationships in Roumania and Poland are good. In particular, he cites the positive experience of transactions which involve the sale of trademark rights along with technology. Such arrangements give the Western firm a vested interest in maintaining the quality of products produced with its technology, since its profits depend on sales of resultant output. At the same time, the Eastern firm must cede to the Western partner an important managerial function – quality control. The Honeywell Inc. agreement with the Polish Mera Union of Automation and Measuring Equipment, discussed by Hayden, the International Harvester Co. agreement with BUMAR, studied by both Hayden and Garland-Marer [19], and the 1971 agreement between Fiat and the Polish government, studied by CEI [7] are apparently successful (with respect to the key goals of both partners) examples of this kind of agreement. Each of the authors, however, describes some implementation problems for these projects.

b) *Effects of mechanisms on assimilation*

Most of the case studies suggest that the continued evolution of technology transfer mechanisms toward more active arrangements is desirable from the perspective of Eastern economic planners because they facilitate more rapid and efficient assimilation of Western technology. The relationship between the type of mechanism and the effectiveness of technology transfer is discussed in a report sponsored by the United States Department of Defense (the so-called "Bucy Report") [72]. The report surveys transfers of four kinds of sophisticated technologies – air frames, jet engines, semiconductors and instrumentation – and concludes that there is a direct, positive relationship between active participation by the transferor and the effectiveness of the technology transfer. Among the highly effective mechanisms were turnkey factories, licenses with extensive teaching effort, joint ventures, technical exchange with ongoing contact, training in high-technology areas, and sales of processing equipment with know-how. Among the more ineffective mechanisms were licenses without accompanying know-how, sales of products, commercial proposals, commercial literature and trade exhibits. The highly effective mechanisms involved extensive and long term interchanges of people between the transferors and recipients, while the ineffective mechanisms involved only passive personal contacts.

Holliday cites cases in the Soviet automotive industry as examples of the Soviet government's willingness to experiment with new technology transfer arrangements, involving greater participation of Western firms in the economy [32]. At the Volga Automobile Plant, for example, he claims that the Soviet government allowed a greater interchange of businessmen, technicians and engineers than in the past. He describes the scientific and technological co-operation agreement signed with Fiat and the extensive training and technical assistance provided by Fiat, and suggests that this may be a prototype for future Soviet arrangements with Western firms. Holliday concludes that such active participation of Western firms is an important precondition to making the technology transfer process more effective and suggests that the Soviet automobile industry is becoming more receptive to such arrangements.

The CIA study of Western technology transfer to the Soviet automotive industry, however, reaches a different conclusion [68]. It emphasizes an important change in Soviet technology transfer mechanisms since the beginning of the Soviet automotive industry. From total dependence on the Ford Motor Company to build the Gorky Automobile Plant in the 1930s, the study suggests that the industry has achieved a state of technological sophistication that obviates the need for active, large-scale participation of Western firms in future Soviet automotive projects. The CIA study maintains that, even at the Volga Automobile Plant and the Kama River Truck Plant, the roles of Western firms were more limited than is commonly perceived. (Western technology transfer to the Soviet automotive industry is the focus of a number of case studies. Holliday studies the Gorky, Volga and Kama River Plants. Sutton also discusses the Gorky and Volga projects. Baranson, Chase World Information [8], and Stowell [90] discuss aspects of the Kama River project. Hanson-Hill's case studies on machine-tools apparently include sales to the Volga and Kama River projects, although they are not clearly identified.)

Goodman [21] presents a case study – of the Soviet computer industry – in which technology transfers have occurred primarily through passive mechanisms, such as reading the Western technical literature or examining Western hardware. He implies that the absence of more active technological ties to the West was one (of several) reasons for a Soviet technological lag in this sector. (Nevertheless, Goodman observes that there have been respectable achievements in the Soviet computer industry.) The reliance on passive mechanisms, he suggests, was partially by choice. Although Western export controls were an important factor, Goodman maintains that Soviet officials did not use the available computer technology transfer sources and mechanisms to the extent that they might have.

A CIA study of the Soviet gas industry [64] suggests that purchases of Western technology for that sector have also relied chiefly on passive mechanisms. The study maintains that Western technical assistance in field development and pipeline construction would be helpful, but that Soviet reluctance to allow on-site Western participation makes this unlikely. The OTA study of the Soviet energy sector [71] also concludes that the assimilation of Western technology in the oil and gas industries has been impeded by Soviet refusal to allow hand-on training by Western suppliers to be carried out in the field.

4. HARD CURRENCY CONSTRAINTS AND EFFORTS TO EXPAND EXPORTS

To some extent, all East European countries' purchases of Western technology are constrained by shortages of hard currency. Generally, the case studies provide evidence that the shortage of hard currency and the need to stimulate hard currency earnings are critical considerations in Eastern technology import decisions. The authors of the case studies devote considerable attention to techniques used by Eastern foreign trade officials to minimise or compensate for outlays of hard currency. Among such techniques are: hard, prolonged bargaining tactics designed to win the best possible terms for the purchase; requests for long term financing at favourable terms; and assistance on compensation or buy-back agreements.

a) *Hard currency expenditures*

Most case studies provide data on hard currency expenditures for individual contracts, and some attempt to aggregate Soviet expenditures on Western machinery by sector. In a general study of Soviet import needs during 1976-1970, for example, Hanson derives aggregate statistics by sector from official Soviet foreign trade statistics [83]. Other studies, such as the CIA studies on the Soviet oil and chemical industries [66, 67], Tasky [59, 60], Grant [22], Hanson [25], and CIA [94], apparently compile sectoral import data by monitoring machinery orders placed with Western countries. OTA uses Western trade statistics to compile Soviet purchases of machinery and equipment for the energy sector [71].

It is frequently observed that total hard currency costs for a project can be much larger than the initial contract value. Total costs can, for example, include purchases of intermediate products or components not anticipated when the contract was signed, or follow-up purchases of improvements in the technology. No effort has been made, however, to compile systematically all such hard currency costs for a major project.

A common question, addressed in the case studies as well as the general literature on East-West trade, is the extent to which Eastern foreign trade officials "whipsaw" Western firms (play one firm off against another in order to bargain down the purchase price) during negotiating sessions. Hanson-Hill report that such tactics had been observed by United Kingdom machine-tool suppliers, who believed that the extended proposal and negotiation times frequently allowed Soviet buyers to receive extremely favourable commercial terms. Marer also reports that United States chemical equipment suppliers suspect such practices [45]. Likewise, Hanson suggests that the Soviets had negotiated good terms on purchases of ammonia plants. (They had paid "certainly a modest amount in relation to returns on ammonia plant".) [25] Hayden, on the other hand, finds no evidence in his six case studies of successful whipsawing. Generally, he concludes, United States firms refused to bow to threats that alternative suppliers existed and succeeded in bargaining for profitable contracts [30]. De Pauw finds cases where the whipsaw technique was used, but notes that it was not always successful [12].

Many authors attest to the desire of Eastern trade officials for long-term financing, which appears to be an integral part of most of the transactions discussed in the case studies. Marer lists the ability to arrange long-term financing on competitive terms as a precondition for United States firms to sell chemical plants to the East. Several authors emphasize the importance of official export credits and provide examples in the automotive and chemical sectors in which the terms offered by official credit institutions appear to have been a major factor in Eastern import decisions [44, 32, 45].

b) *Tying technology purchases to exports of resultant products*

There is general agreement among the authors of the case studies that the desire to expand exports is a major rationale for increased Eastern imports of Western technology. The ECE study of East-West co-operation in the automotive sector [61] observes typically that during the 1965-1974 period, Eastern policymakers sought a rapid expansion of co-operation agreements in order to:

   a) spur domestic industrialisation and modernisation through an expansion of their automotive industries;
   b) meet the increasing domestic consumer demand for passenger cars; and
   c) increase the output of consumer durables which could be sold for convertible currencies in order to finance imports of capital equipment.

While specific comparisons have not been made, the case studies suggest that the emphasis placed on generating exports, as opposed to meeting domestic needs, varies among Eastern countries and industrial sectors. Nevertheless, most of the case studies indicate a substantial degree of interest in generating exports from projects assisted by Western technology. There is also agreement among most of the case studies with the ECE automotive study's conclusion that the Eastern emphasis on exports has grown during the 1970s, with the result of a greater incidence of buy-back agreements and co-production arrangements involving joint marketing in the West. Where no formal contractual obligation for buybacks or joint marketing exists, the case studies suggest that many Western-assisted projects are nonetheless allocating a part of their output for exports to the West.

Thus, McMillan-St. Charles emphasize that balance-of-payments considerations were a major concern for Eastern European officials who decided to allow joint ventures with Western firms [43]. The self-financing features of the joint ventures they study, with Eastern hard currency earnings paying for imports of technology, are extremely attractive to Eastern planners confronted with hard currency constraints. All of the joint ventures which they study involve substantial exports to the West, with some type of assistance from the Western firm's international marketing network. In each case, the Western firm appears to have encouraged and profited from the hard currency sales aspect of the agreement.

Other authors (Levcik-Stankovsky [41], Hayden [30], Garland-Marer [19], Schnitzer [53] and the ECE automotive and machine-tool [63] studies) cite numerous examples of industrial co-operation agreements in which both Western and Eastern partners have vested interests in generating exports from the Western-equipped project. In some cases, such as the Clark Equipment Company agreement with Poland and the Instrument Systems Corporation and Control Data Corporation agreements with Roumania discussed by Hayden, Western firms have actively sought low-cost European production sources and consequently accepted finished products. In such cases, two kinds of conditions – provisions delineating a division of foreign markets and responsibility for quality control – are common. Western firms frequently try to retain exclusive marketing rights in the West as well as some rights in Eastern Europe and developing countries. Eastern partners, on the other hand, usually bargain to gain a share of Western markets. Hayden observes that bargaining leverage and the motivations of the partners often determine whether products from the East are marketed in the West. When the agreement does provide for exports to the West, the Western partner commonly insists on maintaining some control over the quality of output. For example, such a provision was included in the International Harvester-BUMAR agreement, studied by Garland-Marer and the Fiat-Poland agreements, studied by CEI [7].

Agreements including buy-back or compensation provisions are frequently concluded at the insistance of the Eastern partners, who see such provisions as an effective means of penetrating Western markets. A study by the United States International Trade Commission (ITC) [95] concludes that Eastern officials have tended to negotiate countertrade agreements for some types of products more than for others. Generally, Western firms which sell the following types of products encounter Eastern demands for countertrade most often:

- low-priority (as ranked by central government) consumer items, especially types already domestically produced such as television receivers, washing machines, shoes, etc.;
- products not destined for import under current five-year plans;
- large industrial projects normally requiring massive amounts of Western financing, including projects given higher priority in current five-year plans (e.g., chemical projects); and
- industrial projects not employing sophisticated Western technology.

According to the ITC study, exporters of the following items to the East encounter fewer demands for countertrade:

- grains and other foodstuffs;
- machinery, equipment, and manufacturing processes employing sophisticated Western technology (e.g., computers);
- products given high priority under current five-year plans;
- products needed for energy development, such as oil and gas production and processing equipment, certain types of construction machinery used in the building of pipelines and storage facilities; and
- mineral fuels (e.g., coal).

An OECD study of countertrade transactions in East-West trade discusses the kinds of products exported from the East under such arrangements [88]. The study distinguishes between commercial compensation (also called counter-purchases) and industrial compensation (also called buy-back agreements). The former refers to short-term arrangements in which the Eastern exports are unrelated to the imported product. Industrial compensation or buy-back agreements, on the other hand, refer to arrangements which are generally large-scale and long-term and in which the Eastern exports are directly produced by the imported Western technology.

According to the OECD study, the Eastern mechanical engineering industries (machine-tools, industrial vehicles, cranes, etc.) provide the largest share of products supplied as counterpurchases. The best of these products come from the German Democratic Republic, Czechoslovakia and Hungary. Eastern industries producing consumer goods are the second largest source of counterpurchase products. Such products tend to be of low quality and difficult to market in the West. Electrical and electronic equipment and components are sometimes used as counterpurchase products. With the exception of those produced in the German Democratic Republic and Hungary, according to the study, these products tend to be far below Western manufacturing and design standards. Eastern industrial chemicals are also exported under counterpurchase arrangements. They tend to be of satisfactory quality, as Western standards are widely used. There are, however, frequent problems caused by supply breakdowns. The study finds that few raw materials and semi-finished products are involved in counterpurchase arrangements; they are more often exported under industrial compensation transactions. Oil, gas and wood from the Soviet Union, coal, sulphur and copper from Poland, and bauxite from Hungary are prominent examples of products exported under the latter type of transaction.

While many Western firms successfully resist countertrade agreements (Hayden, Levcik-Stankovsky, Schnitzer), some of the case studies indicate that they are becoming more frequent and may result in a substantial increase in Eastern exports of manufactured goods to the West. (ECE automotive study, the OECD study of the chemical industry [49] and the CIA study.) For example, the ECE study estimates that the value of exports to the West under countertrade arrangements in the automotive sector will increase from approximately $.5 billion annually in the late 1970s to $2-3 billion in the early 1980s. The large expected increase is due primarily to major projects that were under construction at the time of the study, such as large plants in the German Democratic Republic and Roumania assisted by Citroën, a co-production venture between Poland and Steyr-Daimler-Puch for trucks, and a joint venture between Roumania and Renault.

An OECD study estimates that compensation agreements signed in the chemical sector in the 1970s will result in annual exports valued at $1.1-1.5 billion in the 1980s (almost 90 per cent of which will come from the Soviet Union) [49]. An earlier CIA study concluded that chemical compensation agreements signed through mid-1977 would result in annual chemical exports to the West of over $300 million. The ITC report concludes that during 1981-1990 annual exports from compensation agreements in the chemical industry would exceed $1.5 billion. The ITC finds that largely as a result of compensation agreements, East European exports of chemicals to the United States had already increased rapidly from $48 million in 1975 to $305 million in 1980.

c) *Competitiveness of Eastern exporters*

The case studies provide varying appraisals of future competitive pressures from Eastern manufacturers. Of them, Baranson [1] offers the most alarmist appraisal, citing examples of East-West industrial co-operation agreements that are likely to result in the export of sophisticated manufactured products from the East. The estimate cited above of future chemical exports from compensation agreements also suggest Western market disruption problems. The OECD, CIA and ITC studies, for example, all conclude that intense competition from Eastern ammonia, produced largely with imported Western plants, is highly likely. The OECD also indicates prospective competitive pressures for soda ash and plastic products. For most chemical products, however, the studies do not anticipate intense competition from the East.

Most of the case studies take a sceptical view of the ability of Eastern enterprises to manufacture products that are competitive on Western markets. A more typical assessment is provided in a Stanford Research Institute study (SRI) [42] of four Soviet industries – semiconductors, commercial aircraft, construction machinery and equipment and synthetic fibres. The study, which attempts to assess the competitive threat posed by transfers of Western technology to those industries, concludes that they are unlikely to pose a major competitive threat to United States firms. At best, the study concludes, Soviet exports are likely to achieve only marginal success in a few product lines. The report cites numerous reasons for this assessment: weaknesses in technological innovation; lack of responsiveness to users' needs; poor maintenance; poor after-sales support services; and inexperience in marketing abroad.

The case studies cite a variety of other factors that impede Eastern export competitiveness. First, many Western firms have negotiated restrictive marketing arrangements. The ECE study on East-West co-operation in the machine-tool sector, for example, found that most agreements were based on a strict allocation of markets, with Western firms retaining exclusive marketing rights in the West for jointly produced machine-tools. The study observes that few Western firms conceded to Eastern demands for marketing rights in the West. Gutman [24] makes the same observation with respect to industrial contracts in the automotive industry. He notes that as a general rule contracts for building turnkey plants, or for manufacturing under license, contain restrictive clauses limiting the risk of large-scale exports.

Another reason for doubting the ability of Eastern enterprises to compete on Western markets lies in the chronic infrastructural problems, discussed below, which have contributed to poor quality standards, technological obsolescence, and delays in startup times and production schedules. Gutman, for example, emphasizes the lack of Western consumer appeal of older model automobiles produced in Eastern Europe under industrial co-operation agreements. Juhasz observes that Hungarian exports of numerically controlled machine-tools have suffered because delivery dates are one-and-a-half to two times as long as West European market averages [36].

The CIA automotive study suggests another reason why Eastern exports of manufactured goods may be limited: Eastern products are often designed for the different conditions that exist in the domestic economies. Thus, the study observes that Soviet trucks and passenger cars are designed for unusual Soviet conditions, such as poor roads, low-octane gasoline, low-grade lubricants and cold weather. It suggests that Soviet vehicles are not well suited for operating conditions in the West and, therefore, are unlikely to be exported in large quantities. Moreover, several studies indicate that for many industrial sectors, recipients of Western technology have far to go just to meet domestic demand. They suggest that there will not be substantial surpluses of many products available for export. The OECD and CIA chemical studies, for example, find that this is the case for many chemicals being produced with Western technology.

Effective assimilation of Western technology is generally seen as a precondition to expansion of exports to Western markets. For manufactured exports, Western technology is seen as a key to achieving the quality levels needed to compete in the West. Thus, Tasky's study of the East European computer industries [59] and the CIA study of the chemical industry suggest that the assimilative capacity of the Eastern economies for Western technology will be an important determinant of future Eastern exports in those sectors. The SRI study concludes that poor Soviet performance in absorbing and maintaining advanced technology is likely to be a major impediment to Soviet export competitiveness in the four industries it surveys.

The formidable barriers confronting Eastern planners in their efforts to expand exports of manufactured goods to the West suggest that traditional Eastern exports, particularly raw materials, will continue to play a major role in East-West trade. Studies by the CIA [64, 65, 66], OTA [71] and the United States Congressional Joint Economic Committee [70] of the Soviet energy sector highlight the importance of Soviet oil and gas exports to Soviet hard currency earnings. All of the studies suggest that natural gas exports are likely to replace oil exports as the major hard currency earner for the Soviet Union. They emphasize, however, that Western technology may be a key to expanding production and generating larger exportable surpluses.

5. INDICATORS OF ASSIMILATIVE CAPACITY

While assimilative capacity is a difficult concept to quantify, the authors of some case studies have provided data which indicate roughly the degree of success in assimilating foreign technologies. Specifically, they have provided information on lead-times, operation of imported plants after start-up and, to a lesser degree, the rate of diffusion.

a) *Lead-times*

Lead-times (defined by Hanson-Hill as the time taken to acquire, install and begin operating imported equipment) provide one indicator of the assimilative capacities of the Eastern economies. Most existing case studies which analyse lead-times conclude that Western-assisted projects are chronically slow in beginning operation. For example, an ECE survey of Western firms involved in industrial co-operation in the machine-tool sector [63] found that all firms reported that a one-year delay in starting production and meeting export obligations was customary. At least three firms reported delays of up to four years before the projects were functioning smoothly. Marer [45] observes that negotiations between Western firms and officials in the Eastern chemical industries require too much time and expense. He adds that performance test runs normally take several weeks, or months, compared with periods of one week or less in the United States. De Pauw also found that negotiations in the United States-Soviet transactions he studied were unusually lengthy [12]. The CIA automotive study and Holliday state that startup has taken substantially longer than planned at major Western-assisted projects. The CIA chemical industry study notes that while startup of Western-equipped chemical plants in the Soviet Union may be slow (3-5 years after orders are placed), they are considerably faster than domestic startup times (which average eight years).

Lead-times are the focus of the Hanson-Hill and Röthlingshöfer-Vogel surveys. They examine systematically the time required for various stages of the technology transfer process – from initial negotiations to startup – and attempt to make comparisons with times required for startup of similar plants in Western industrial countries. In the Hanson-Hill survey, most of the companies supplying machine-tools commented on the lengthy time interval between the initial inquiry and the final agreement on a proposal. In several cases, this took more than two years – about three times longer than for West European customers. The survey indicated that the longest delays occurred in the installation and commissioning stage, which often required about three times as long as those stages in Western industrial countries. They conclude that overall, lead-times in the Soviet Union are probably between two and three times the expected time spans for a factory in the West. Similarly, they conclude that 26 turnkey chemical plant transactions in the survey took an average of six years, 10 months – about three-and-a-half to four times longer than West European plants normally require.

Röthlingshöfer-Vogel's survey indicates that negotiations require twice as long in transactions with Soviet purchasers than with customers in the West. Subsequent stages were also more time-consuming. In both surveys, respondents observed that long lead-times were the result of Soviet bureaucratic procedures (especially communication delays within the foreign trade organisation and between the foreign trade organisation and the end-users), Soviet negotiating techniques, and other infrastructural problems.

b) *Operation after startup*

The case studies generally conclude that assimilation problems have not only delayed startup times, but have also adversely affected operation of plant and equipment after startup. For example, some of the respondents in the Hanson-Hill and Röthlingshöfer-Vogel surveys give rough comparisons of manning levels and output with equivalent plants in the West. Hanson-Hill provide estimates for 13 of the chemical projects included in their survey. The average manning levels of the plants in the East exceeded West European levels by 50-70 per cent. Nine of the plants were determined to be operating at output levels lower than would be expected in the West; three at about the same level; and one contract (relating to four plants) was believed to be generating output at levels above similar plants in Western Europe. Röthlingshöfer-Vogel found that the level of manning was an estimated 20 per cent higher than for similar plants in the West and that the rate of capacity utilisation was about 80 per cent of Western levels. They speculate that labour productivity at the Eastern plants was about two-thirds of what would be expected in the West.

Inefficient operation of imported plants appears to be widespread in the East, and it is described in a number of case studies. The inefficiency of Eastern assimilation, however, is relative: performance is considered poor by comparison with Western industrial countries. When compared with existing Eastern plants, Western-equipped plants generally represent substantial improvements in efficiency. While extensive, reliable data are apparently not available, Eastern published sources and Western observers have provided useful comparative information.

The CIA chemical study concludes that the Soviet chemical industry has made substantial gains in overall efficiency and product quality as a result of technology imports, but that the gains have come more slowly than anticipated by Soviet planners. The study provides a specific example of improvements in efficiency in Western-equipped plants. It notes that, despite poor management, Western-supplied ammonia plants have substantially cut Soviet production costs: the plants use 5 per cent of the electric power and one-third fewer workers than the most recent Soviet-designed plants. During 1971-1975, average production costs for ammonia fell 8.5 per cent as a result of new installations. As noted above, Hanson [25] makes similar observations about improvements in the efficiency of Western-equipped mineral fertilizer plants and in the general productivity of the agricultural sector. Garland-Marer [19], Bojko [3], Baranson [1], CEI [7] and others also describe improvements in productivity and product quality at individual Western-assisted projects.

c) *Diffusion*

One of the important ways in which technology transfer contributes to improved economic performance is through diffusion of the imported technology. Diffusion pertains to the extent to which new technologies are duplicated by other domestic enterprises and become assimilated throughout the economy. As noted above, it is generally considered to have a decisive influence on the rate of technological progress in the industry as a whole. (See Amann [76] for further discussion.) Despite its importance, diffusion has received little attention in the existing case studies.

The Hanson-Hill and Röthlingshöfer-Vogel surveys provide sketchy information on diffusion. Hanson-Hill report that, while relatively few respondents had information, there was a general impression among chemical equipment suppliers that the Soviets had not acquired the capability to duplicate imported plants. Some respondents noted instances of successful copying of individual items, while others noted instances of unsuccessful attempts to copy entire plants. Röthlingshöfer-Vogel found only one company that had reason to believe that Soviet specialists had attempted to duplicate certain technical principles of the plants it had supplied. Similarly, the CIA chemical study concludes that the Soviets have not been able to copy Western chemical equipment on a substantial scale. It adds, however, that they probably can produce spare parts for some plants and apparently were successful in negotiating for the right to re-use Western designs for some parts of large ammonia plants.

The CIA automotive study provides evidence to the contrary on the rate of diffusion of imported technology in the Soviet Union. The study concludes that "Imported equipment has been readily assimilated into Soviet plants and copied by the Soviet automotive and machine-tool industries" [68]. This statement is atypical of the findings of most observers who have studied diffusion in the Soviet economy.

Young, for example, finds little evidence of diffusion of the massive amounts of Western technology purchased for the Volga River Automobile Plant and the Kama River Truck Plant [74]. On the contrary, he finds a tendency for other Soviet automobile plants to buy similar technologies from the same Western suppliers. Young's analysis suggests that the Soviet pattern of importing entire production complexes facilitates the initial adoption of Western technology, but may impede diffusion to other parts of the economy. Most other detailed analyses of diffusion in Soviet industry conclude that it has been relatively slow, although some exceptions have been noted. (Amann [76] and Slama-Vogel [54].)

6. EFFECTS OF DOMESTIC INFRASTRUCTURES AND INSTITUTIONS ON ASSIMILATIVE CAPACITY

Those case studies which focus on the division of labour between Western transferor firms and recipient enterprises in the East tend to agree that any successful transfer of technology requires major complementary inputs from the recipient's domestic economy. Thus, Holliday discusses the high priority assigned to major Western-assisted projects and describes how scarce resources (high-quality materials, skilled labour, and technical expertise) are sometimes diverted from other parts of the economy in order to put the projects into operation. He thus suggests that the assimilation of Western technology has been very costly in terms of domestic resources. Similarly, the CIA automotive study emphasizes that most inputs for such large Western-assisted automotive projects as the Volga Automobile Plant and the Kama River Truck Plant come from domestic resources. Hewett [31] and OTA describe the enormous domestic inputs required to complement Western machinery and equipment for large Soviet energy projects.

Because of the large inputs required from the domestic economy, such case studies emphasize the importance of the domestic infrastructure to the success of the technology transfer process. Even when the Western firm supplies a complete plant with extensive managerial and technical assistance, the speed and efficiency of the assimilation process depend heavily on the quality of domestic design and construction organisations, complementary industries, the supply system, engineers, managers and workers. In addition, a number of case studies suggest that the assimilative capacity of a country depends heavily on the economic and political environment it provides for technology imports.

Gustafson maintains that the assimilative capacity of an industry in the recipient country depends primarily on the technological level the industry had attained before the import of Western technology [23]. The Soviet Union, he finds, has been able to benefit most from technology imports to those sectors which were already operating at a relatively high technological level. Thus, Gustafson finds that the Soviet chemical industry continues to lag behind its Western counterparts despite massive imports of Western technology because of the lack of indigenous research and development capabilities. Soviet computer technology also lags behind Western standards because of domestic infrastructural and systemic problems. On the other hand, Gustafson finds that Western automotive technology has contributed significantly to the Soviet industry largely because the Soviet Union already possessed considerable capabilities in that sector.

a) *Factors which impede assimilation*

A pervasive theme of many case studies is that weaknesses in the Eastern economic and technological infrastructures reduce the capacity to assimilate Western technology rapidly and efficiently. The surveys by Röthlingshöfer-Vogel and Hanson-Hill itemise a number of infrastructural problems in the Soviet Union which tend to slow down and make less efficient the assimilative capacity of the economy. Both cite deficiencies in the construction sector (poor planning, equipment and materials); low skills of workers, engineers and managers in the construction work, installation and plant operation; and poor co-ordination of plans and investments in related plants and industries, resulting in inadequate supplies. Röthlingshöfer-Vogel also note the absence of adequate research and development facilities: although there are high-quality specialists in central research institutes, they are separated from the production process and lack the necessary facilities to carry out research and development on a large scale.

Such infrastructural weaknesses are cited in a number of other case studies. For example, the CIA study on the Soviet chemical industry echoes the description of deficiencies in Soviet research institutes. It also notes that the rapid rate of technology imports has created shortages of labour and materials, thus possibly limiting the amount of further technology imports for the chemical sector. Holliday describes serious construction and supply problems for new Soviet automotive projects, and Hayden's case studies include complaints of worker apathy and lack of technical skills.

Several studies emphasize systemic features which impede effective assimilation. An SRI study of technology transfer to the Soviet Union, for example, cites the lack of managerial incentives to innovate, the separation of research and development from the production process, barriers to travel to and from the Soviet Union, reluctance to agree to active technology transfer arrangements, and the absence of competitive pressures on Soviet enterprises [42]. Systemic barriers to effective assimilation are also emphasized in other case studies. Hanson-Hill find complaints of slow decision-making, lack of entrepreneurship, lack of incentives, and resistance to innovation. Le Bihan [39] and Radice [50] note the lack of co-ordination and poor communications because of the inability of Western firms to deal directly with independent end-users. M.J. Berry, in a study of innovation in the Soviet machine-tool industry, emphasizes the Soviet industry's lack of information about technological developments in the West (which have been significant in recent years). He concludes that the lack of information has impeded Soviet importers' ability to make intelligent choices on which technologies to import [77].

b) *Factors which facilitate assimilation*

Although weaknesses tend to be emphasized, observations in the case studies on Eastern infrastructures are not all negative. Hayden, while pointing out problems in the assimilation process, concludes that none were insurmountable. In all cases, he finds, the recipients possessed the necessary general technology; that is, all had the basic production capability for the new products and processes that were introduced. Garland-Marer praise the entrepreneurial qualities of Polish BUMAR officials. They also attribute to BUMAR officials considerable experience in manufacturing, good engineering design, modern equipment, and capable and well-trained machinists. Hanson-Hill, while emphasizing weaknesses in the infrastructure of the Soviet chemical industry, note the existence of some dynamic and competent officials.

Of the case studies surveyed, the CIA study of the Soviet automotive industry presents the most favourable assessment of Soviet assimilative capacity. It maintains that the technological level of Soviet automotive engineering is high, and that, consequently, it is able to utilise advanced technology effectively. The study acknowledges that the Soviet automotive industry has been indifferent toward innovation, but maintains that the lag in Soviet automotive engineering stems from economic factors (such as central planning, lack of incentives, long production runs on standardised projects and professional isolation) rather than inferior technological competence. Despite such economic impediments, the study argues, the Soviet automotive industry has carried out significant technological development independently of the West. (Thus, the author of the study is apparently drawing a distinction between the technological competence of Soviet engineers, which he maintains is high, and the actual conditions of large parts of the industry, which he suggests is not abreast with the leading Western industries because of systemic impediments.)

Hanson-Hill also note an important Soviet systemic feature – the ability to overcome problems by mobilising domestic resources (especially labour) needed to complete projects – which contributes positively to the assimilative capacity of the Soviet chemical industry. Holliday describes a similar phenomenon at major Soviet automotive projects. The Volga Automobile Plant and the Kama River Plant, he observes, were assigned high priority in allocation of such technological resources as high quality machinery and materials and talented engineers and managers. Holliday maintains that the ability to channel resources is a major contributing factor to the considerable Soviet achievements in mastering foreign technology.

III. CONCLUSIONS AND SUGGESTIONS FOR FUTURE CASE STUDIES

1. CONCLUSIONS ON ASSIMILATIVE CAPACITY AND IMPACT ON TRADE

While a considerable number of case studies are cited in this survey, it must be emphasized that relatively few of them provide detailed and systematic analyses of the central questions related to the assimilative capacity of the Eastern economies and the impact of technology transfers on East-West trade. Indeed, many of the studies deal with technology transfer only peripherally: they are primarily assessments of various industries in the Eastern countries or analyses of prospective markets for Western firms. Thus, the number of relevant and informative case studies is small.

Moreover, the existing studies are somewhat tentative in their conclusions about questions of assimilation and trade impact. Most importantly, they suggest that generalisations about these questions should be made with caution. One could use the findings of the case studies to support the proposition that Eastern projects most often assimilate Western technology slowly and inefficiently in comparison with projects in the industrial West. In some cases, however, the case studies have provided evidence of relatively rapid and efficient assimilation. One could conclude that the Eastern need for Western technology will continue to grow, creating an expanding market for Western capital goods and other products. However, the case studies provide at least some evidence to the contrary. One could also conclude that most Eastern industrial sectors have not excelled at manufacturing goods that are competitive on Western markets. The case studies, however, point to possible competitive threats in some product lines.

Despite the tentative nature of their conclusions, it may be useful to summarise briefly the major findings that are suggested by the majority of the authors of the existing case studies.

a) *Level of technology and degree of dependence on the West.* Western technology has made a significant contribution to the technological advancement of the Eastern industrial branches studied. Those branches are likely to continue to rely on imports of Western technology. Eastern importers insist on receiving advanced, up-to-date technologies, but, for a number of reasons, the most recent technologies are not transferred. Eastern importers are increasingly emphasizing contractual arrangements which provide for continuing updates of the imported technologies, as opposed to simple, one-time transfers.

b) *Channels or mechanisms of technology transfer.* During the 1960s and 1970s, there was an evolution toward long-term "active" technology transfer mechanisms in East-West trade. Such arrangements facilitate the more rapid and efficient assimilation of technology in the Eastern economies. There are still, however, limitations on the degree of participation by Western firms and the exchange of personnel in the technology transfer process.

c) *Hard currency constraints and efforts to expand exports.* Hard currency shortages and the need to stimulate exports to the West are critical considerations in Eastern technology import decisions. An increasing number of technology transfer contracts include provisions for counterpurchases or buy-backs of products manufactured with the imported technologies. Eastern projects benefiting from Western technology imports, however, have had limited success in competing on Western markets.

d) *Indicators of assimilative capacity.* When compared with the experiences of Western industrial countries, assimilation of imported technologies has been slow and inefficient in the East.

e) *Effects of domestic infrastructure and institutions on assimilative capacity.* The assimilative capacity of Eastern industries depends heavily on the quality of complementary inputs from the domestic economy. Shortcomings in the technological infrastructures and the political and economic institutions of the Eastern countries tend to inhibit rapid and efficient assimilation of Western technology and the export of competitive manufactured goods to the West.

It should be re-emphasized that there are significant differences among the authors of existing studies. While foreign technology-assisted projects in the East may tend to perform poorly relative to comparable projects in the West, there are enough examples of successful technology assimilation and export penetration of Western markets to make extreme conclusions unwise.

The findings of the case studies are largely in agreement with findings of much of the general literature on East-West technology transfer. They are, for example, mostly consistent with the findings of three OECD studies on Western transfer of technology to three Eastern European countries – Poland, Czechoslovakia, and Roumania. (Fallenbuchl [81]; Levcik-Skolka [85]; and Kaser [84]. Additional country studies are forthcoming.) Like the sectoral case studies, the OECD country studies find that the Eastern countries have generally poor assimilative capacities for Western technology and that Eastern efforts to penetrate Western markets have met with only limited success. Also like the sectoral case studies, the country studies attribute poor Eastern performance to infrastructural and systemic shortcomings.

There is one significant discrepancy between the findings of the sectoral case studies and findings of the more general literature, including the country case studies. The sectoral case studies tend to emphasize that, despite problems of assimilation, Western technology has made a major contribution to technological progress and growth of those sectors studied. The country studies, on the other hand, find little evidence that Western technology has had a major beneficial impact on the Eastern economies as a whole. This discrepancy can be explained largely by the different approaches of the studies. Eastern planners have allocated a disproportionate share of their scarce hard currency reserves to certain high priority sectors. The sectoral studies are concentrated on those Eastern sectors which have been the primary recipients of Western technology. By comparing imports of the favoured sector with total output for the sector, they naturally find that the contribution is relatively large. The country studies, on the other hand, use an essentially macro-economic approach. They compare imports of Western technology with total economic activity in the country, and find the contribution of Western technology relatively small. The different findings of the country and sectoral case studies may also be partially due to important methodological problems in measuring the contribution of technology imports to economic growth [82].

Do the case studies provide criteria for determining the conditions under which Western technology is likely to be assimilated rapidly and have a significant effect on

East-West trade? They certainly do not provide definitive criteria. The case studies do, however, provide considerable evidence to support several hypotheses about the determinants of assimilative capacity and export competitiveness. There is substantial evidence, for example, to suggest that the following criteria may be used to determine the assimilative capacity of Eastern industry.

1. The technological level of the recipient industry or sector is the primary determinant of assimilative capacity.
2. The kinds of mechanisms used to transfer technology largely determine the outcome. Active mechanisms facilitate the rapid and efficient assimilation.
3. The political and economic systems of the Eastern countries are major impediments to assimilation. Without significant reform, there is little prospect of improving assimilative capacities.

Under what conditions are Eastern industries likely to stimulate the production of goods that are competitive on Western markets? Two hypotheses are particularly noteworthy:

1. The key to more successful Eastern export performance is an improvement in assimilative capacity. Assimilation of Western technology is a prerequisite for penetration of Western markets.
2. A greater involvement of Western firms in quality control and marketing assistance is the best means of improving export performance.

There is no overwhelming "proof" for any of these hypotheses. Various case studies have, however, presented substantial evidence for each of them. They deserve further attention and testing in future case studies.

While the existing case studies have provided useful insights, they leave major gaps in our understanding of assimilation and trade impact which should be addressed in future case studies. The existing literature contains extensive discussions of such important questions as the mechanisms of technology transfer, the technological levels of Eastern industries, and the interaction of Eastern and Western firms in the pre-commissioning stages of plant construction. What is needed is a more thorough analysis of other key questions: the assimilation process after the first startup of a new technology; the extent and quality of domestic complementary inputs for Western-assisted projects; the effect of technology transfers on future Western exports to the East; and the degree to which Eastern exporters have succeeded in meeting necessary quality standards for Western markets and the techniques they have used to accomplish that end.

With respect to assimilation, the Hanson-Hill and Röthlingshöfer-Vogel studies have provided excellent analyses of initial lead-times in selected Soviet industries. As noted in the introduction, however, it is performance in the other phases of assimilation – lead-times for successive imports of a given technology and the extent and speed of diffusion – which largely determine the general technological advancement and growth of an industry. Analyses of improvements in the assimilation process would require case studies of extended periods of time, probably 10 years or more. Effective techniques for studying the speed and extent of diffusion have been developed [54, 76], but have not been widely utilised in the case studies. For most industries, there is evidence on the dates of first commercial production of a new product or the first application of a new process. A useful indicator of the extent of diffusion is the percentage of output produced by a new process at a given time (see Hanson [26]).

The efficiency of the assimilation process in the East has also been inadequately explored. To what extent are scarce technological resources wasted during the technology transfer process and after the startup of an imported technology? If foreign exchange and domestic resource scarcities are a serious constraint on assimilation of foreign technology, as suggested above, the total costs of assimilating technology deserve more attention. The extent and nature of domestic expenditures, in particular, should be a focus of future case studies.

Important questions about the effects of Western technology transfer to the East on trade (in both directions) have not received adequate attention in existing case studies. With respect to Western exports, for example, does the output produced with imported machinery and equipment substitute for traditional Eastern imports? Does the import of technology generate additional demand for imports of complementary and intermediate goods and updated technology? With respect to Eastern exports to the West, much has been written about the export orientation of various Western technology-assisted projects. A useful focus for future case studies would be an examination of the degree of success Eastern firms have had in meeting export goals or commitments. Some of the estimates that have been made of future counter-deliveries, for example, assume that startup schedules will be met and planned output levels will be achieved. It would be useful to test such assumptions in cases where counter-deliveries have already begun.

2. SUGGESTED METHODOLOGIES AND SOURCES FOR FUTURE CASE STUDIES

Case studies of technology transfers from West to East are mostly of recent vintage: the vast majority have been published since 1977. Consequently, the methodologies and approaches are to some degree experimental. Nevertheless, they provide some excellent examples from which authors of future case studies might learn or in some cases emulate. In particular, they provide guidelines on how to contend with a chronic problem of students of the Eastern economies – the paucity of reliable information and data. To be sure, none of the existing case studies completely solves this problem. The combination of official secrecy in the East and corporate secrecy in the West is a formidable barrier. Some authors of case studies, however, have succeeded in skillfully exploiting the limited sources that exist, thus providing useful insights into the questions of assimilative capacity and trade impact.

The problem of protecting proprietary information while still providing useful data and information about technology transfers has been a common one for authors of the existing case studies. It is a particularly difficult problem in studying individual projects. (It is not, however, an insoluble one, as evidenced by the Garland-Marer study.) The existence of such an information barrier suggests that a focus on Western technology transfers to Eastern branches or sectors, as opposed to individual projects, is a more fruitful approach for future case studies. For example, Western corporate executives are likely to be more candid in answering questionnaires if details of individual contracts can be disguised in more general case studies. (This was the experience of the Hanson-Hill and Röthlingshöfer-Vogel studies.)

Hanson and Hill demonstrate convincingly that certain aspects of assimilative capacity – lead-times, output and manning levels of imported plant when it is in operation, and the rate of diffusion – are amenable to study by surveying the experiences of Western exporters. Their case studies (and the similar ones by Röthlingshöfer-Vogel) provide the best data available on lead-times, as well as limited, but useful information on the other questions they investigated. A major strength of their approach is the use of standardised and carefully drafted questionnaires which provide a comparable approach for each of the case studies. Such an approach appears to generate data which can be more readily assimilated and analysed than data obtained from informal interviews. The authors acknowledge some shortcomings in their approach. The Western exporters' knowledge of the operation of the technology after startup was limited. The authors' identification and description of individual projects were constrained by the companies' desire to avoid disclosure of trade secrets or proprietary information. Many companies declined to participate. (Still, the authors believe that their sample is representative in most respects.) Despite these shortcomings, the Hanson-Hill and Röthlingshöfer-Vogel case studies illustrate how future case studies can benefit by carefully examining this source of information.

De Pauw's study of commercial negotiations in United States-Soviet trade, focusing on Control Data Corporation's experience, also successfully obtained valuable information about the negotiation stage. Like the Hanson-Hill and Röthlingshöfer-Vogel studies, De Pauw's study benefited from the use of standardised questionnaires, followed by extensive interviews of corporate officials.

Other case studies indicate that interviews with Western exporters provide detailed information on contracts and the technical characteristics of products and processes that are not available elsewhere. Of particular interest are countertrade provisions and joint marketing arrangements, knowledge of which can provide a sound basis for discussing the impact of technology transfers on future East-West trade. Several of the case studies included in this survey obtained such data from Western corporate sources. There is, however, no evidence of a systematic survey, similar to those discussed above, of the trade impact of various technology transfers from West to East.

An ECE survey which focuses on the trade impact of East-West industrial co-operation in the automotive industry notes that few Western executives were prepared to discuss details of counter-deliveries. Consequently, the study relied heavily on published sources for such information. Nevertheless, it seems likely that questionnaires could be designed to solicit useful information on the trade impact of such transactions, while protecting proprietary information. For example, questions could be posed concerning the value of counter-deliveries as a percentage of the original purchase price and problems in meeting quality standards and delivery schedules. Standardised questionnaires including questions on export arrangements should be a part of future case studies.

All of the case studies suggest that certain Western published sources provide information that is complementary to that which can be obtained in interviews. For example, various newsletters and periodicals monitor Eastern plant and equipment orders and provide some data on contract terms. Other sources provide supplemental information on the experience of Western firms transferring technology to the East.

Perhaps the most underutilised sources of information and data are Soviet and East European publications. Eastern published sources sometimes provide information on the operation of Western-assisted projects after startup, diffusion of new technologies and other economic and technological developments in industry. The specialised technical literature in various Eastern countries tends to be more candid and informative than the general press in discussing industrial developments. Polish and Hungarian publications, in particular, provide serious analyses of the problems of assimilating foreign technology and producing exports for Western markets. (A survey of the East European and Soviet literature is outside the scope of this paper. Fallenbuchl [81] cites examples of Polish studies on technology transfer, and Abonyi [75] examples of Hungarian studies. See also Kiss [37], Elias [16], Tardos [91] and Juhasz [36] for examples of the Hungarian perspective.) Generally speaking, it appears that authors of existing case studies have not fully exploited the Eastern literature on technology transfer.

Although trade statistics do not provide fully satisfactory measures of the transfer of technology, they are an important element of most useful case studies. Drabek and

Slater [80] note the difficulties of using trade statistics alone to draw firm conclusions about the relationship between technology imports and exports. Eastern export performance, they point out, is a function of other factors, such as changing market conditions and indigenous technological developments. Nevertheless, their compilation of Eastern industrial goods exports to the West by branch of origin and of Eastern capital goods imports from the West by industrial end-use branch provides a crude basis for quantifying the relationship between technology imports and exports. Drabek-Slater also suggest that the statistics they have compiled are amenable to exploitation in more commodity detail and on a country-by-country basis in a manner which could explore more clearly trends in individual commodity and country performance. They make a convincing case that examination of the trade data is "an indispensable first step" in defining the relationship between Eastern technology imports and exports and in identifying sectors for further scrutiny.

Authors of future case studies should make some effort to quantify the contribution of Western technology to the industry they are studying. While the relationship between inputs of Western technology and final output is a complex one, some rough indicators are possible. Hanson's study of the Soviet mineral fertilizer industry [25] provides a useful model for future case studies.

Ideally, all sources of information – interviews, technical and trade journals, and official statistical data – should be exploited in future case studies. Used together, they are likely to produce a more comprehensive picture than is currently available of the assimilative capacities and export performance of key Eastern industrial sectors.

3. SUGGESTED SECTORS FOR FUTURE CASE STUDIES

The most interesting subjects for future sectoral case studies are those sectors that have been major recipients of Western technology and have had, or are likely to have, a significant impact on East-West trade. Two likely candidates, the chemical and automotive industries, have already received substantial attention in case studies, which provide a useful base of information. Nevertheless, careful exploitation of the data and information that are available could provide many more insights into the questions of assimilation and trade in those sectors. For most of the Eastern countries, these two industries have received high (in some cases, the highest) priority for technology imports, and many observers anticipate significant Eastern exports from Western-assisted projects. Much remains to be studied at the subsectoral level, that is, individual projects or product lines. Moreover, the lengthy history of technology transfers to Eastern automotive and chemical industries provides an excellent opportunity to study possible changes in assimilation and trade performance over time.

Generally, the export orientation of the automotive and chemical industries would be a useful focus for future case studies. Both industries contain projects with a significant portion of output allocated for export to Western markets. A particularly interesting question about the two industries is the possible emergence of a pattern of Eastern specialisation, producing basic chemicals or automotive sub-assemblies which could be exported to pay for imported Western technology.

In addition, future studies should emphasize the experience of the automotive and chemical industries in the smaller East European countries. The existing literature is concentrated on the Soviet Union. For example, the Polish (chemical and passenger car) and Hungarian (chemical and public transportation vehicle) industries may prove to be useful case studies because of their prominence in countertrade arrangements and the likelihood of finding information and data in domestic publications. The automotive and chemical industries of the German Democratic Republic and Roumania also deserve consideration because of their size and their experience as recipients of Western technology.

Several other sectors, which have received some attention in the existing literature, deserve further analysis because their development has benefited from important inputs of Western technology and they have had or may have significant effects on East-West trade. The electronics (especially computer) industries of most of the Eastern countries are prominent examples. Czechoslovakia, which has relied heavily on imports of Western computers, the German Democratic Republic, which has relied mostly on development of an indigenous industry, and Roumania, which has emphasized importation of computer manufacturing technology, provide interesting contrasts. A comparative analysis of the three industries would be a useful case study. The Hungarian pharmaceuticals industry, traditionally an important branch of the economy and an important source of internationally traded pharmaceuticals, is another interesting subject for a case study. Hungarian and Polish manufacturers of agricultural and construction equipment are also important recipients of Western technology which deserve consideration for future case studies.

Few case studies have focused on technology transfers to raw materials development projects in the East, even though such transfers and Eastern exports of raw materials account for a major share of East-West trade. These include sectors, such as the Soviet oil, gas and timber industries and the Polish coal industry, that produce products with great demand in the West. The scarcity of good case studies in these sectors may reflect unusually severe problems in collecting suitable data. (Soviet publications and statistics, for example, may provide less information on raw materials sectors, and Western businessmen and technicians probably have less first-hand experience with them than with other industries.) Nevertheless, the importance of Eastern raw materials output for East-West trade suggests a need to assign them high priority in future case studies.

Questions about the effects of technology imports on Soviet oil and gas output and exports have assumed special significance because of major political and strategic, as well as economic, implications. Negotiations between the Soviet Union and several Western countries over construction of a major new pipeline, for example, have highlighted the growing importance of natural gas exports in Soviet trade with the West. The pipeline may result in a major increase in Soviet imports of Western machinery and equipment and in Soviet hard currency earnings. The size and importance of this project suggest that it should be considered a high priority for a future case study.

Finally, two industries – shipping and textiles – which are important beneficiaries of Western technology transfers have received little attention in the existing literature. Western technology appears to have contributed to the success of Eastern shipping lines in competing with Western shipping firms. It may thus be an interesting example of successful assimilation and penetration of Western markets. The Eastern textiles industries present an interesting contrast. Although significant importers of Western technology, they do not appear to have posed a competitive threat on Western markets. Studies of the contribution of Western technology to these two industries, and the reasons for their different export roles should provide useful insights into the questions of assimilation and trade.

In conclusion, the authors of future case studies can draw fruitfully from the existing literature. It provides examples of useful methodologies or approaches for case studies, identifies important sources of information, and reaches interesting, if tentative, conclusions about Western technology transfer to various Eastern industries. However, the existing studies represent only an initial effort, and further case studies should add substantially to our understanding of the assimilative capacities of the Eastern economies and the impact on trade of West-East technology transfers.

NOTES

1. Eastern Europe, as used in this paper, refers to Bulgaria, Czechoslovakia, the German Democratic Republic, Hungary, Poland and Roumania.
2. The author is indebted to Dr. Philip Hanson for suggesting the various aspects of assimilation.
3. Numbers refer to items cited in the Bibliography.
4. A detailed discussion of the various forms of technology transfer is beyond the scope of this paper. See Zaleski-Wienert [96], ECE [92], McMillan [86], and Levcik-Stankovsky [41] for further analysis.

IV. SELECTED BIBLIOGRAPHY

1. CASE STUDIES

[1] Baranson, Jack, *International Transfer of Industrial Technology by US Firms and their Implication for the US Economy,* US Department of Labor, International Labor Affairs Bureau, Office of Foreign Economic Research, Washington, D.C., 24th September, 1976.

[2] Bauer, T. and Soos, K.A., "Inter-Firm Relations and Technological Change in Eastern Europe – The Case of the Hungarian Motor Industry", *Acta Oeconomica,* Vol. 23, Nos. 3-4, 1979, pp. 285-303.

[3] Bojko, Bela, "Results and Problems of Co-operation with Western Firms in Hungarian Light Industry", in C.T. Saunders, ed., *East-West Co-operation in Business: Inter-Firm Studies,* Springer-Verlag, Vienna, 1977.

[4] Brada, Josef C., "Industry Structure and East-West Technology Transfer: A Case Study of the Pharmaceutical Industry", *The ACES Bulletin,* Vol. XXII, No. 1 (Spring 1980), pp. 31-59.

[5] Campbell, Robert, *Soviet Energy Technologies,* Indiana University Press, Bloomington, 1980.

[6] Campbell, Robert, "Technology Transfer in the Energy Sector: East-West and Global Issues", in C.T. Saunders, ed., *Industrial Policies and Technology Transfer Between East and West,* Springer-Verlag, Vienna, 1977.

[7] Centre d'études industrielles, *Fiat in Poland – A Case Study,* by Michael Katz, Geneva, 1979.

[8] Chase World Information Corporation, *KamAZ, The Billion Dollar Beginning,* New York, 1974.

[9] Cheburakov, "East-West Commercial and Industrial Co-operation in Machine-Tool Building", in C.T. Saunders, ed., *East-West Co-operation in Business: Inter-Firm Studies, op. cit.*

[10] Crane, Keith and Gilliot, Antoine, "MNCs in the Aircraft Industries in Brazil and Poland", Paper prepared for Conference on Multinational Corporations in Latin America and Eastern Europe, Bloomington, Indiana, 5th-8th March, 1981.

[11] Davis, N.C. and Goodman, S.E., "The Soviet Bloc's Unified System of Computers", *ACM Computing Surveys,* June 1978.

[12] De Pauw, John W., *Soviet-American Trade Negotiations,* Praeger, New York, 1979.

[13] Economist Intelligence Unit, *Rubber and the Automotive Industry in the USSR,* Special Report No. 28, London, December 1976.

[14] Edwards, Imogene U., "Automotive Trends in the USSR", in US Congress Joint Economic Committee, *Soviet Economic Prospects for the Seventies,* 93rd Congress, 1st session, Washington, D.C., 27th June, 1973, pp. 291-314.

[15] Edwards, Imogene U., and Fraser, Robert, "The Internationalisation of the East European Automotive Industries", in US Congress, Joint Economic Committee, *East European Economies Post-Helsinki,* 95th Congress, 1st session, Washington, D.C., 25th August, 1977, pp. 396-419.

[16] Elias, Andras, "The Application of Western Technology in the Hungarian Agriculture and Food Industries", *The ACES Bulletin,* Vol. XXII, No. 1 (Spring 1980), pp. 61-82.

[17] Eronen, Jarmo, "L'industrie des pâtes et papiers en Union soviétique: Bilan et perspectives", *Le courrier des pays de l'Est,* No. 259, February 1982, pp. 51-67.

[18] Garland, John, "US MNCs and the Construction Equipment Industry in Latin America and Eastern Europe", Paper prepared for the International Conference on Multinational Corporations in Latin America and Eastern Europe, Bloomington, Indiana, 5th-8th March, 1981.

[19] Garland, John and Marer, Paul, "US Multinationals in Poland: A Case Study of the International Harvester-BUMAR Co-operation in Construction Machinery", in US Congress, Joint Economic Committee, *East European Economic Assessment,* Part I: *Country Studies, 1980,* 97th Congress, 1st session, Washington, D.C., 27th February, 1981, pp. 121-137.

[20] Gèze, François, "La coopération Est-Ouest dans l'industrie électronique", *Le courrier des pays de l'Est*, June 1979, pp. 3-36.

[21] Goodman, Seymour E., "Soviet Computing and Technology Transfer: An Overview", *World Politics*, No. 4, July 1979, pp. 539-570.

[22] Grant, James, "Soviet Machine-Tools: Lagging Technology and Rising Imports", in US Congress, Joint Economic Committee, *Soviet Economy in a Time of Change*, 96th Congress, 1st session, Washington, D.C., 10th October, 1979, pp. 554-580.

[23] Gustafson, Thane, *Selling the Russians the Rope? Soviet Technology Policy and United States Export Controls*, Rand Corporation, Santa Monica, April 1981.

[24] Gutman, Patrick, "Coopération industrielle Est-Ouest dans l'automobile et modalités d'insertion des pays de l'Est dans la division internationale du travail occidentale", *Revue d'études comparatives Est-Ouest*, June-September, 1980, (translated as "East-West Industrial Co-operation in the Automotive Industry and the International Division of Labour", *The ACES Bulletin*, Vol. XXII, Nos. 3-4 (Fall-Winter 1980), pp. 1-64.

[25] Hanson, Philip, "The Impact of Western Technology: A Case Study of the Soviet Mineral Fertilizer Industry", in Paul Marer and John Michael Montias, eds., *East European Integration and East-West Trade*, Indiana University Press, Bloomington, 1980.

[26] Hanson, Philip, *Trade and Technology in Soviet-Western Relations*, Columbia University Press, New York, 1981.

[27] Hanson, Philip and Hill, Malcolm R., *Soviet Absorption of Western Technology: A Survey of West European Experience*, (Report on the Survey of UK Exporters), Stanford Research Institute, December 1978.

[28] Hardt, John P. and Holliday, George, D., "Technology Transfer and Change in the Soviet Economic System", in Frederic J. Fleron, ed., *Technology and Communist Culture*, Praeger, New York, 1977.

[29] Hardt, Rolf, "Looking Back at Ten Years' Co-operation of a West German Firm with Polish Machine-Tool Builders", in Saunders, C.T., ed., *East-West Co-operation in Business: Inter-Firm Studies, op. cit.*

[30] Hayden, Eric W., *Technology Transfer to Eastern Europe: US Corporate Experience*, Praeger, New York, 1976.

[31] Hewett, Edward A., "Near-Term Prospects in the Soviet Gas Industry and the Implications for East-West Trade", in US Congress, Joint Economic Committee, *Soviet Economy in the 1980s: Problems and Prospects*, Part 1, Washington, D.C., 1983.

[32] Holliday, George D., *Technology Transfer to the USSR, 1928-1937 and 1966-1975: The Role of Western Technology in Soviet Economic Development*, Westview Press, Boulder, Colorado, 1979.

[33] Improta, Bruno and Lungo, R., "Experience of a West European Company in the Chemical Field", in Saunders, C.T., ed., *Industrial Policies and Technology Transfers Between East and West, op. cit.*

[34] Internationales Zentrum für Ost-West Kooperation, GmbH and Chase World Information Corporation, *KFZ-und KFZ-Teileindustrie in den RGW Ländern und Jugoslawien*, June 1979.

[35] Judy, Richard W., "The Case of Computer Technology", in Wasowski, Stanislaw, ed., *East-West Trade and the Technology Gap*, Praeger, New York, 1970.

[36] Juhasz, Jeno, "Some Aspects of the Adaptation of the Most Advanced Technical Achievements in Hungary", *The ACES Bulletin*, Vol. XXII, No. 1 (Spring 1980).

[37] Kiss, Eva Eszter, "Problems of Technology in the Hungarian Pharmaceutical Industry", *The ACES Bulletin*, Vol. XXII, No. 1 (Spring 1980).

[38] Kosnik, Joseph T., *Natural Gas Imports from the Soviet Union*, Praeger, New York, 1975.

[39] Le Bihan, Joseph, "East-West Co-operation in Agri-business", in Saunders, C.T., ed., *East-West Co-operation in Business: Inter-Firm Studies, op. cit.*

[40] Lent, Harold, "East European Chemical Production and Trade", in US Congress, Joint Economic Committee, *Reorientation and Commercial Relations of the Economies of Eastern Europe*, 93rd Congress, 2nd session, Washington, D.C., 16th August, 1974, pp. 394-405.

[41] Levcik, Friedrich and Stankovsky, Jan, *Industrial Co-operation Between East and West*, M.E. Sharpe, Inc., White Plains, New York, 1978.

[42] Levine, Herbert S.; Movit, Charles H.; Earle, Mark M., Jr. and Liebermann, Anne R., *Transfer of US Technology to the Soviet Union: Impact on US Commercial Interests*, prepared for the US Department of State, Stanford Research Institute, Strategic Studies Center, SRI Project 3543, February 1976.

[43] McMillan, C.H. and St. Charles, D.P., *Joint Ventures in East Europe: A Three-Country Comparison*, The Canadian Economic Policy Committee, C.D. Howe Research Institute, Canada, 1974.

[44] Malzacher, H. Michael, "Practical Aspects of East-West Co-operation Projects for an Austrian Firm", in Saunders, C.T. ed., *East-West Co-operation in Business: Inter-Firm Studies, op. cit.*

[45] Marer, Paul, "US-CMEA Industrial Co-operation in the Chemical Industry" in Saunders, C.T. ed., *East-West Co-operation in Business: Inter-Firm Studies, op. cit.*

[46] Müller, Friedmann, *Produktions- und Aussenwirtschaftsbeziehungen der RGW-Länder im Energiesektor bis 1990 – Voraussetzungen einer Ost-West-Zusammenarbeit*, Stiftung Wissenschaft und Politik, Ebenhausen, February 1979.

[47] Mundie, D.A. and Goodman, S.E., *The Integration of the Comecon Computer Industries*, A Report prepared for the National Council for Soviet and East European Research, Washington, D.C., July 1981.

[48] Nykryn Jaroslav, "Inter-firm Co-operation in the Czechoslovak Machine Building Industry", in Saunders, C.T., ed., *East-West Co-operation in Business: Inter-Firm Studies, op. cit.*

[49] OECD, *East-West Trade in Chemicals*, Paris, 1980.

[50] Radice, Hugo, "Experiences of East-West Industrial Co-operation: A Case Study of UK Firms in the Electronics, Telecommunications and Precision Engineering Industries", in Saunders, C.T., ed., *East-West Co-operation in Business: Inter-Firm Studies, op. cit.*

[51] Röthlingshöfer, Karl Ch. and Vogel, Heinrich, *Soviet Absorption of Western Technology*, Report for the Stanford Research Institute, March 1979.

[52] Schaum, Jack H., "KamAZ Foundary...USA on Display", *Modern Casting*, March 1976, pp. 42-55.

[53] Schnitzer, Martin, *US Business Involvement in Eastern Europe: Case Studies of Hungary, Poland and Roumania*, Praeger, New York, 1980.

[54] Slama, Jiri and Vogel, Heinrich, "Zur Verbreitung neuer Technologien in der USSR – eine Fallstudie: das Oxygenblasstahlverfahren", *Jahrbücher für Nationalökonomie und Statistik*, Vol. 187, No. 3, 1973.

[55] Sobeslavsky, V. and Beazley, P., *The Transfer of Technology to Socialist Countries: The Case of the Soviet Chemical Industry*, Oelgeschlager, Gunn and Hain, Publishers, Inc., Cambridge, Mass., 1980.

[56] Sobral, N. and Hinks-Edwards, M., *The Communist Bloc Automobile Industry: Implications for Western Manufacturers*, Euroeconomics, 1975.

[57] Stein, John Picard, *Estimating the Market for Computers in the Soviet Union and Eastern Europe*, Rand Corporation, Santa Monica, 1974.

[58] Sutton, Antony C., *Western Technology and Soviet Economic Development*, Volume I: *1917-1930*, Volume II: *1930-1945*, Volume III: *1945-1964*, Hoover Institution Press, Stanford, 1968, 1971 and 1973.

[59] Tasky, Kenneth, "Eastern Europe: Trends in Imports of Western Computer Equipment and Technology", in US Congress, Joint Economic Committee, *East European Economic Assessment*, Part II: *Regional Assessments*, 97th Congress, 1st session, Washington, D.C., 10th July, 1981, pp. 296-327.

[60] Tasky, Kenneth, "Soviet Technology Gap and Dependence on the West: The Case of Computers", in US Congress, Joint Economic Committee, *Soviet Economy in a Time of Change*, 96th Congress, 1st session, Washington, D.C., 10th October, 1979, pp. 510-523.

[61] United Nations Economic Commission for Europe, *East-West Co-operation in the Automotive Sector and Counter Trade Arrangements*, TRADE/R.385/Add. 1, Geneva, 8th October, 1979.

[62] United Nations Economic Commission for Europe, *Promotion of Trade Through Industrial Co-operation. Case Studies in Industrial Co-operation. Results of a Survey of Five Western Enterprises*, TRADE/R.373/Add. 3, Geneva, 1978.

[63] United Nations Economic Commission for Europe, *Promotion of Trade Through Industrial Co-operation: The Experience of Selected Western Enterprises Engaging in East-West Industrial Co-operation. Results of a Survey of Fifteen Firms in the Machine-Tool Sector*, TRADE/R.373/Add. 4, Geneva, 1978.

[64] US Central Intelligence Agency, National Foreign Assessment Center, *USSR: Development of the Gas Industry*, ER 78-10393, Washington, D.C., July 1978.

[65] US Central Intelligence Agency, National Foreign Assessment Center, *Prospects for Soviet Oil Production*, ER 77-10270, Washington, D.C., April 1977.

[66] US Central Intelligence Agency, National Foreign Assessment Center, *Prospects for Soviet Oil Production- A Supplemental Analysis*, ER 77-10425, Washington, D.C., July 1977.

[67] US Central Intelligence Agency, National Foreign Assessment Center, *Soviet Chemical Equipment Purchases from the West: Impact on Production and Foreign Trade*, ER 78-10554, Washington, D.C., October 1978.

[68] US Central Intelligence Agency, National Foreign Assessment Center, *USSR: Role of Foreign Technology in the Development of the Motor Vehicle Industry*, ER 79-10571, Washington, D.C., October 1979.

[69] US Congress House Committee on Banking and Currency, Subcommittee on International Trade, *The Fiat-Soviet Auto Plant and Communist Economic Reforms*, 89th Congress, 2nd session, Washington, D.C., 1st March, 1967.

[70] US Congress, Joint Economic Committee, Subcommittee on International Trade, Finance and Security Economics, *Energy in Soviet Policy*, 97th Congress, 1st session, Washington, D.C., 11th June, 1981.

[71] US Congress, Office of Technology Assessment, *Technology and Soviet Energy Availability*, Washington, D.C., 1981.

[72] US Department of Defense, Office of the Director of Defense Research and Engineering, *An Analysis of Export Control of US Technology – A DOD Perspective*, A Report of the Defense Science Board Task Force on Export of US Technology, Washington, D.C., 4th February, 1976.

[73] Welihozkiy, Toli, "Automobiles and the Soviet Consumer", in US Congress, Joint Economic Committee, *Soviet Economy in a Time of Change*, 96th Congress, 1st session, Washington, D.C., 10th October, 1979, pp. 811-833.

[74] Young, John P., *Impact of Soviet Ministry Management Practices on the Assimilation of Imported Process Technology – With Examples From the Motor Vehicle Sector*, Paper presented at the Joint Annual Meeting of the Southwestern and Rocky Mountain Associations of Slavic Studies, Houston, Texas, 13th April, 1978.

2. OTHER SOURCES CITED

[75] Abonyi, Arpad, "Imported Technology, Hungarian Industrial Development and Factors Impeding the Emergence of Innovative Capability", in Radice, Hugo; Hare, Paul and Swain, Nigel, eds., *Economic Trends and Economic Management in Hungary*, Allen and Unwin, London, 1981.

[76] Amann, Ronald; Cooper, Julian and Davies, R.W., eds., *The Technological Level of Soviet Industry*, Yale University Press, New Haven, 1977.

[77] Amann, Ronald and Cooper, Julien, *Industrial Innovation in the Soviet Union*, Yale University Press, New Haven, 1982.

[78] Bolz, Klaus and Plötz, Peter, *Erfahrungen aus der Ost-West Kooperation*, HWWA-Institut für Wirtschaftsforschung, Verlag Weltarchiv, GmbH, Hamburg, 1974.

[79] Drabek, Zdenek, *Western Embodied Technology and its Sectoral Impact on East European Exports to the West*, Unpublished paper, November 1981.

[80] Drabek, Zdenek and Slater, John, *East-West Technology Transfer: Flows of Technology and Technology-Related Products*, Unpublished paper.

[81] Fallenbuchl, Zbigniew M., *East-West Technology Transfer. Study of Poland 1971-1980*, OECD, Paris, 1983.

[82] Gomulka, Stanislaw and Nove, Alec, "Econometric Evaluation of the Contribution of West-East Technology Transfer to the East's Economic Growth". Part I of the present volume.

[83] Hanson, Philip, *USSR: Foreign Trade Implications of the 1976-1980 Plan*, Economist Intelligence Unit, Special Report No. 36, London, October 1976.

[84] Kaser, Michael, *East-West Technology Transfer: The Flow of Technology from OECD Members to Roumania*, Unpublished paper.

[85] Levcik, Friedrich and Skolka, Jiri, *East-West Technology Transfer. Study of Czechoslovakia*, OECD, Paris, 1984.

[86] McMillan, Carl H., "East-West Industrial Co-operation", in US Congress, Joint Economic Committee, *East-European Economies Post-Helsinki*, 95th Congress, 1st session, Washington, D.C., 25th August, 1977, pp. 1175-1224.

[87] Marer, Paul and Miller, Joseph C., "US Participation in East-West Industrial Co-operation Agreements", *Journal of International Business Studies*, Vol. 8, No. 2 (Fall-Winter 1977), pp. 17-29.

[88] OECD, *East-West Trade: Recent Developments in Countertrade,* Paris, 1981.

[89] Slama, Jiri and Walter, F.J., *Zur Frage der Auswirkungen des Technologietransfer West-Ost auf das Wirtschaftwachstum der UdSSR,* Industrieanlagen-Betriebsgessellschaft, 1980.

[90] Stowell, Christopher E., *Soviet Industrial Import Priorities, with Marketing Considerations for Exporting to the USSR,* Praeger, New York, 1975.

[91] Tardos, Marton M., "Importing Western Technology into Hungary", in Bornstein, Morris; Gitelman, Zvi and Zimmerman, William, *East-West Relations and the Future of Eastern Europe: Politics and Economics,* Allen and Unwin, London, 1981.

[92] United Nations, Economic Commission for Europe, *Analytical Report on Industrial Co-operation among ECE Countries,* Geneva, 1973.

[93] US Central Intelligence Agency, *Soviet Acquisition of Western Technology,* in "Remarks of William L. Armstrong", *Congressional Record* (daily ed.), Vol. 128, Washington, D.C., 19th May, 1982, pp. S5589-S5594.

[94] US Central Intelligence Agency, National Foreign Assessment Center, *The Soviet Economy in 1978-79 and Prospects for 1980,* ER 80-10328, June 1980.

[95] US International Trade Commission, *Analysis of Recent Trends in US Countertrade,* Washington, D.C., March 1982.

[96] Zaleski, Eugène and Wienert, Helgard, *Technology Transfer Between East and West, OECD,* Paris, 1980.

67587DEPO1.3S-

OECD SALES AGENTS
DÉPOSITAIRES DES PUBLICATIONS DE L'OCDE

ARGENTINA – ARGENTINE
Carlos Hirsch S.R.L., Florida 165, 4° Piso (Galería Guemes)
1333 BUENOS AIRES, Tel. 33.1787.2391 y 30.7122

AUSTRALIA – AUSTRALIE
Australia and New Zealand Book Company Pty. Ltd.,
10 Aquatic Drive, Frenchs Forest, N.S.W. 2086
P.O. Box 459, BROOKVALE, N.S.W. 2100

AUSTRIA – AUTRICHE
OECD Publications and Information Center
4 Simrockstrasse 5300 Bonn (Germany). Tel. (0228) 21.60.45
Local Agent/Agent local :
Gerold and Co., Graben 31, WIEN 1. Tel. 52.22.35

BELGIUM – BELGIQUE
Jean De Lannoy, Service Publications OCDE
avenue du Roi 202, B-1060 BRUXELLES. Tel. 02/538.51.69

BRAZIL – BRÉSIL
Mestre Jou S.A., Rua Guaipá 518,
Caixa Postal 24090, 05089 SAO PAULO 10. Tel. 261.1920
Rua Senador Dantas 19 s/205-6, RIO DE JANEIRO GB.
Tel. 232.07.32

CANADA
Renouf Publishing Company Limited,
2182 ouest, rue Ste-Catherine,
MONTRÉAL, Qué. H3H 1M7. Tel. (514)937.3519
OTTAWA, Ont. K1P 5A6, 61 Sparks Street

DENMARK – DANEMARK
Munksgaard Export and Subscription Service
35, Nørre Søgade
DK 1370 KØBENHAVN K. Tel. +45.1.12.85.70

FINLAND – FINLANDE
Akateeminen Kirjakauppa
Keskuskatu 1, 00100 HELSINKI 10. Tel. 65.11.22

FRANCE
Bureau des Publications de l'OCDE,
2 rue André-Pascal, 75775 PARIS CEDEX 16. Tel. (1) 524.81.67
Principal correspondant :
13602 AIX-EN-PROVENCE : Librairie de l'Université.
Tel. 26.18.08

GERMANY – ALLEMAGNE
OECD Publications and Information Center
4 Simrockstrasse 5300 BONN Tel. (0228) 21.60.45

GREECE – GRÈCE
Librairie Kauffmann, 28 rue du Stade,
ATHÈNES 132. Tel. 322.21.60

HONG-KONG
Government Information Services,
Publications/Sales Section, Baskerville House,
2nd Floor, 22 Ice House Street

ICELAND – ISLANDE
Snaebjörn Jönsson and Co. h.f.,
Hafnarstraeti 4 and 9, P.O.B. 1131, REYKJAVIK.
Tel. 13133/14281/11936

INDIA – INDE
Oxford Book and Stationery Co. :
NEW DELHI-1, Scindia House. Tel. 45896
CALCUTTA 700016, 17 Park Street. Tel. 240832

INDONESIA – INDONÉSIE
PDIN-LIPI, P.O. Box 3065/JKT., JAKARTA. Tel. 583467

IRELAND – IRLANDE
TDC Publishers – Library Suppliers
12 North Frederick Street, DUBLIN 1 Tel. 744835-749677

ITALY – ITALIE
Libreria Commissionaria Sansoni :
Via Lamarmora 45, 50121 FIRENZE. Tel. 579751/584468
Via Bartolini 29, 20155 MILANO. Tel. 365083
Sub-depositari :
Ugo Tassi
Via A. Farnese 28, 00192 ROMA. Tel. 310590
Editrice e Libreria Herder,
Piazza Montecitorio 120, 00186 ROMA. Tel. 6794628
Costantino Ercolano, Via Generale Orsini 46, 80132 NAPOLI. Tel. 405210
Libreria Hoepli, Via Hoepli 5, 20121 MILANO. Tel. 865446
Libreria Scientifica, Dott. Lucio de Biasio "Aeiou"
Via Meravigli 16, 20123 MILANO Tel. 807679
Libreria Zanichelli
Piazza Galvani 1/A, 40124 Bologna Tel. 237389
Libreria Lattes, Via Garibaldi 3, 10122 TORINO. Tel. 519274
La diffusione delle edizioni OCSE è inoltre assicurata dalle migliori librerie nelle città più importanti.

JAPAN – JAPON
OECD Publications and Information Center.
Landic Akasaka Bldg., 2-3-4 Akasaka,
Minato-ku, TOKYO 107 Tel. 586.2016

KOREA – CORÉE
Pan Korea Book Corporation,
P.O. Box n° 101 Kwangwhamun, SÉOUL. Tel. 72.7369

LEBANON – LIBAN
Documenta Scientifica/Redico,
Edison Building, Bliss Street, P.O. Box 5641, BEIRUT.
Tel. 354429 – 344425

MALAYSIA – MALAISIE
University of Malaya Co-operative Bookshop Ltd.
P.O. Box 1127, Jalan Pantai Baru
KUALA LUMPUR. Tel. 51425, 54058, 54361

THE NETHERLANDS – PAYS-BAS
Staatsuitgeverij, Verzendboekhandel,
Chr. Plantijnstraat 1 Postbus 20014
2500 EA S-GRAVENHAGE. Tel. nr. 070.789911
Voor bestellingen: Tel. 070.789208

NEW ZEALAND – NOUVELLE-ZÉLANDE
Publications Section,
Government Printing Office Bookshops:
AUCKLAND: Retail Bookshop: 25 Rutland Street,
Mail Orders: 85 Beach Road, Private Bag C.P.O.
HAMILTON: Retail: Ward Street,
Mail Orders, P.O. Box 857
WELLINGTON: Retail: Mulgrave Street (Head Office),
Cubacade World Trade Centre
Mail Orders: Private Bag
CHRISTCHURCH: Retail: 159 Hereford Street,
Mail Orders: Private Bag
DUNEDIN: Retail: Princes Street
Mail Order: P.O. Box 1104

NORWAY – NORVÈGE
J.G. TANUM A/S
P.O. Box 1177 Sentrum OSLO 1. Tel. (02) 80.12.60

PAKISTAN
Mirza Book Agency, 65 Shahrah Quaid-E-Azam, LAHORE 3.
Tel. 66839

PHILIPPINES
National Book Store, Inc.
Library Services Division, P.O. Box 1934, MANILA.
Tel. Nos. 49.43.06 to 09, 40.53.45, 49.45.12

PORTUGAL
Livraria Portugal, Rua do Carmo 70-74,
1117 LISBOA CODEX. Tel. 360582/3

SINGAPORE – SINGAPOUR
Information Publications Pte Ltd.
Pei-Fu Industrial Building,
24 New Industrial Road N° 02-06
SINGAPORE 1953. Tel. 2831786, 2831798

SPAIN – ESPAGNE
Mundi-Prensa Libros, S.A.
Castelló 37, Apartado 1223, MADRID-1. Tel. 275.46.55
Libreria Bosch, Ronda Universidad 11, BARCELONA 7.
Tel. 317.53.08, 317.53.58

SWEDEN – SUÈDE
AB CE Fritzes Kungl Hovbokhandel,
Box 16 356, S 103 27 STH, Regeringsgatan 12,
DS STOCKHOLM. Tel. 08/23.89.00
Subscription Agency/Abonnements:
Wennergren-Williams AB,
Box 13004, S104 25 STOCKHOLM.
Tel. 08/54.12.00

SWITZERLAND – SUISSE
OECD Publications and Information Center
4 Simrockstrasse 5300 BONN (Germany). Tel. (0228) 21.60.45
Local Agents/Agents locaux
Librairie Payot, 6 rue Grenus, 1211 GENÈVE 11. Tel. 022.31.89.50

TAIWAN – FORMOSE
Good Faith Worldwide Int'l Co., Ltd.
9th floor, No. 118, Sec. 2,
Chung Hsiao E. Road
TAIPEI. Tel. 391.7396/391.7397

THAILAND – THAILANDE
Suksit Siam Co., Ltd., 1715 Rama IV Rd.,
Samyan, BANGKOK 5. Tel. 2511630

TURKEY – TURQUIE
Kültur Yayinlari Is-Turk Ltd. Sti.
Atatürk Bulvari No : 191/Kat. 21
Kavaklidere/ANKARA. Tel. 17 02 66
Dolmabahce Cad. No : 29
BESIKTAS/ISTANBUL. Tel. 60 71 88

UNITED KINGDOM – ROYAUME-UNI
H.M. Stationery Office.
P.O.B. 276, LONDON SW8 5DT.
(postal orders only)
Telephone orders: (01) 622.3316, or
49 High Holborn, LONDON WC1V 6 HB (personal callers)
Branches at: EDINBURGH, BIRMINGHAM, BRISTOL,
MANCHESTER, BELFAST.

UNITED STATES OF AMERICA – ÉTATS-UNIS
OECD Publications and Information Center, Suite 1207,
1750 Pennsylvania Ave., N.W. WASHINGTON, D.C.20006 – 4582
Tel. (202) 724.1857

VENEZUELA
Libreria del Este, Avda. F. Miranda 52, Edificio Galipan,
CARACAS 106. Tel. 32.23.01/33.26.04/31.58.38

YUGOSLAVIA – YOUGOSLAVIE
Jugoslovenska Knjiga, Knez Mihajlova 2, P.O.B. 36, BEOGRAD.
Tel. 621.992

Les commandes provenant de pays où l'OCDE n'a pas encore désigné de dépositaire peuvent être adressées à :
OCDE, Bureau des Publications, 2, rue André-Pascal, 75775 PARIS CEDEX 16.
Orders and inquiries from countries where sales agents have not yet been appointed may be sent to:
OECD, Publications Office, 2, rue André-Pascal, 75775 PARIS CEDEX 16.

67587-05-1984

OECD PUBLICATIONS, 2, rue André-Pascal, 75775 PARIS CEDEX 16 - No. 42799 1984
PRINTED IN FRANCE
(92 84 02 1) ISBN 92-64-12565-5